Life, Death and Statistics

Civil registration, censuses and the work of the General Register Office, 1836–1952

Life, Death and Statistics

Civil registration, censuses and the work of the General Register Office, 1836–1952

EDWARD HIGGS

A Local Population Studies Supplement

Published in 2004 by
LOCAL POPULATION STUDIES
Hatfield, Hertfordshire

©Local Population Studies
Department of Humanities, University of Hertfordshire,
Hertfordshire AL10 9AB

ISBN 0 9541621 0 2

Typeset by Cambrian Typesetters, Frimley
Printed by Halstan & Co Ltd, Amersham

Contents

List of Tables and Figures	vi
Preface	vii

1. Origins: civil registration in early Victorian England — 1

2. The genesis of state medical statistics — 22

3. The expansion of the GRO's statistical functions in the High Victorian period — 45

4. Late Victorian medical statistics in an age of inertia — 90

5. 1900-1914: eugenics and the GRO's Indian summer — 129

6. State medical statistics, the dawn of computing and the Edwardian mathematical revolution — 156

7. 1914-1951: registration and statistics in total war and total welfare — 186

8. Conclusions — 216

Appendix 1: *Registrar General's annual reports, Statistical reviews*, and *Decennial supplements*, 1838-1951 — 221

Consolidated Bibliography — 229

Index — 251

List of Tables and Figures

Tables

Table 3:1: GRO Staffing 1840-1921
Table 3:2: Searches in the registers at the central GRO, 1845-1895
Table 3:3: Salaries in the GRO and other departments, derived from the Civil Service estimates for 1861/2.
Table 4:1: Size of the Civil Service
Table 7:1: Staff of the GRO as of 1 April 1939

Figures

Figure 3:1 Delays in publishing the Annual Reports of the Registrar General
Figure 3:2 Pages of text in the Annual Report of the Registrar General
Figure 4:1 Clerical staffing of the GRO
Figure 7:1 Pages of text in the Statistical Review

Preface

The form of this book – an organisational history of a government department – is a rather old fashioned one, and as such requires some explanation. I undertook the research for *Life, Death and Statistics* in the 1990s as part of a project funded by the Wellcome Trust. This had the objective of showing how the official collection of information on people in England in the modern period was part of a larger movement to incorporate the working classes into the nation state as citizens. Rather than collecting information for the purposes of repressive social control, state data gathering was used to give rights, if of a circumscribed nature, to citizens. This argument was to act as a critique of the belief that information was collected for the purposes of direct social control, or as part of a strategy to expand the 'biopower' at the State's command. I intended to do this via a study of the General Register Office (GRO), which administered the civil registration of births, marriages and deaths from 1837, and organised the decennial censuses from 1841 onwards. However, it became clear that such a grandiose project could not be based on the work of a single government department, whatever its intrinsic interest. I therefore put the detailed history of the institution to one side whilst I worked on the much larger project.

After the completion of a work addressing these wider issues[1], I was encouraged by colleagues at the University of Essex, especially Kevin Schürer and Matthew Woollard, to revise my earlier manuscript to make it serve the narrower, although still important function, of illuminating the statistical output of the GRO. Besides collecting information via the civil registration system, censuses, and various surveys, the GRO also published reports summarising this material on a weekly, quarterly, annual and decennial basis. This data is of crucial importance since much of the demographic, medical, economic and

---

[1] Higgs, *The Information State in England*.

social history of England and Wales in the nineteenth and twentieth centuries is based upon these reports. The history of the period would be immensely impoverished if we did not know the size and distribution of the population; the trends in life expectancies and the diseases from which people died; the occupational structure of the workforce; the impact of birth rates on demographic trends and on the position of women; and so on, and so on. The administrative history of the GRO is not, therefore, an end in itself but will, hopefully, lead to a more informed use by historians of the Office's manuscript returns and statistical output.

The idea that the history of an institution explains, or at least constrains, what it does, or produces, can be derived from two main sources. One is at the cutting edge of much modern historiography, and the other represents a somewhat older tradition. The first reason for believing that a history of the GRO would be useful to historians lies in the belief that knowledge of the world is socially constructed – knowledge can be explained as much by the way in which people see the world, as by what actually exists in it. This manner of understanding knowledge springs from numerous sources, from post-structuralism, the Marxist concept of ideology, the theory of meaning in linguistics and of scientific revolutions, and so on.[2] Put in simple terms this comes down in the present case to understanding the output of a body such as the GRO in terms of the aims and objectives of its officers, and in the way in which they understood the world. They asked certain questions in their censuses and surveys because there were certain subjects they were interested in, and they arranged the replies they got in certain ways because it 'made sense' to them. Understanding how the GRO came about; who staffed it; and the outside pressures, and internal constraints, under which it worked, all help us to understand the nature, limitations, and opportunities presented by its published output. The more traditional reason for writing an institutional history is based on the old archival principle of

---

[2] For a short, useful discussion of these ideas see: Jordanova, 'The social construction of medical knowledge'.

administrative provenance – that records are as they are because of the nature and activities of the bodies that produced them. One can only understand the nature and arrangement of documents, including published reports, by tracing the processes that created them. Perhaps the latter concept merely recapitulates the former but without the jargon.

An analysis of the development of the GRO would certainly explain many of the oddities of the GRO's data collection and output. Some of the issues to be addressed include: why civil registration was not in practice compulsory until 1874; why there were two differing series of *Annual reports of the Registrar General* in the mid-Victorian period; why the GRO failed to measure morbidity until the mid-twentieth century; why the Office refused to expand the census questions in the nineteenth century; why the GRO introduced socio-economic classification into the 1911 census but then failed to use it in data analysis in the inter-war period; why the size of the GRO's publications contracted in the late nineteenth century; and so on. None of these matters is inconsequential for an informed use of the published output of the GRO. The present work will concentrate on the civil registration and medical work of the GRO, rather than upon the taking of the census. This partly reflects the extent of the commentary that already exists on the census.[3] But it is also based on an appreciation of the rather anomalous nature of census taking in the history of the GRO in the years under consideration. Until the inter-war period, the census was merely a temporary, if onerous, distraction from the Office's ongoing statistical activities. If one is to understand the development of the GRO, therefore, it is much more important to look at is registration and public health functions.

An administrative history of the GRO that placed its published output in context would plainly be of use to local, demographic historians interested in the population characteristics and vital trends revealed by censuses and civil registration data. However, a number of other historical disciplines will also

---

[3] See, for example: Higgs, *A clearer sense of the census*.

find the development of the GRO of interest. The history of the GRO is at the heart of the development of the discipline of medical statistics in this country and abroad. For example, the international classification of diseases used today had its origins in the medical nosologies first used in the Office in the mid-nineteenth century. Similarly, the development of the statistical, computational, and data management methodologies used by the GRO are important for an understanding of the history of mathematics and of information technology. Thus, the GRO was the first organisation in Britain to use some of the technologies that were the precursors of the modern IT revolution. Lastly, the history of the Office illustrates some major themes in the formation of the modern British state, including the shifting balance between local and central government, and the consolidation of political control within Whitehall. Each of these themes might justify the production of the present book in its own right.

A knowledgeable critic might ask, however, why such a book was necessary when there are a number of works that already cover various aspects of the history of the GRO. I would reply that such works are more or less deficient, and that local historians, historical demographers, medical and other historians do not have a single, authoritative guide to the development of this key government body. The only existing institutional history of the GRO is Muriel Nissel's *People count*.[4] This is a useful book but is hardly an academic work, and does not place the Office in its wider intellectual, administrative, or political context. Similarly, John Eyler's *Victorian social medicine*[5] is very good on the ideas of William Farr, the GRO's Superintendent of Statistics from 1839 to 1880, but fails to fully explore the broader institutional context within which Farr worked. Eyler's emphasis on Farr has also encouraged subsequent scholars to write about the GRO largely in terms of the activities of its chief statisticians, which has led to numerous misunderstandings. The aims, objectives and abilities of the Office's heads, the

---

[4] Nissel, *People count*.
[5] Eyler, *Victorian social medicine*.

Registrar Generals, were just as important for understanding the history of the institution, if not more so.

Simon Szreter's monumental *Fertility, class and gender in Britain* has a great deal on the history of the GRO, leading up to the introduction of the classification of households into socio-economic groupings in the 1911 fertility census. The story he tells is extremely insightful and provocative, and his arguments on the genesis of the system of socio-economic groupings carry great conviction. However, I think that he has been somewhat misled by his desire to see the GRO as a bastion of environmentalism struggling against eugenics. Szreter over-emphasises the importance of this struggle to the GRO, and underestimates the centrality of eugenic ideas to the GRO's own work. Also, by stopping in 1911 he fails to note the limited use the Office subsequently made of its new classification system, and the reasons for this. I have also written on various aspects of the Office's census and civil registration work in a number of books and articles, and *Life, death and statistics* allows me to bring this material together, and to publish new research findings.

The order of the chapters in *Life, death and statistics* is broadly chronological – a reflection of the central concern with the way in which the nature of the GRO's work and output was the outcome of an historical process. Chapter 1 examines the origins of the civil registration system and of the GRO in the early nineteenth century movement to improve legal title to property. How this impacted on the nature of the GRO's statistical output is also examined. The next chapter then explains how a statistical function developed within the Office. This was initially for actuarial work, and the chapter reveals the key role played in this process by Edwin Chadwick and by the first Registrar General, Thomas Lister. Chapter 3 follows the expansion of the Statistical Department of the GRO in the High Victorian period. Whilst recognising the important contribution made to the Office's statistical work by William Farr, the chapter also emphasises the demands placed on it by other government departments, and the administrative genius of the second Registrar General, Major George Graham, in this expansion.

By contrast, Chapter 4 attempts to explain the relative decline of the GRO's statistical project in the last two decades of the nineteenth century. This was due to a number of complex factors but it probably had less to do with changes in intellectual preoccupations, as some have argued, and more to do with administrative problems and the less than stirring leadership of the GRO's third Registrar General, Sir Brydges Henniker. This can be seen in the speed with which the GRO recovered after Henniker's retirement in 1900, the subject of Chapter 5. Some historians have placed this recovery in the context of the rivalry between eugenics and an environmental approach to public health, as championed by the GRO, although this is shown to be something of a misunderstanding.

The final two chapters, one technical and one administrative, examine the key changes to the work of the GRO in the first half of the twentieth century. Chapter 6 looks at the impact of new forms of machine tabulation and statistical theory on the work of the Office. Chapter 7 discusses the re-orientation of the GRO from an independent crusader for local public health reform, to a statistical spear-carrier for central government. The final chapter then attempts to draw out some general themes in the history of the institution. The book concludes with an appendix listing the GRO's non-census publication prior to 1952.

I owe a debt of gratitude to numerous friends and colleagues who have helped to form my ideas about the work of the GRO. Members of the Wellcome Unit for the History of Medicine at the University of Oxford, and of the History Departments of the Universities of Exeter and Essex, have been especially patient with my statistical enthusiasms. Others who deserve special thanks include, Margo Anderson, Martin Campbell-Kelly, Paul Laxton, Eileen Magnello, Graham Mooney, Maggie Pelling, Andrea Tanner, and Simon Szreter. I am indebted to Kevin Schürer and Matthew Woollard for encouraging me to publish this volume via *Local Population Studies*, and I am grateful to the other members of the editorial board of the journal, who took on what is after all a very specialist work. Both the Wellcome Trust and the Leverhulme Trust provided grants to undertake the research upon which this book was based. As ever, I must thank my wife Liz for her support, and for the sensible advice that I should have taken in the first place.

1

Origins: civil registration in early Victorian England

The 1836 Registration and Marriage Acts

Civil registration is a right of passage into the modern state. When parents register the births of their children they place them on the road to formal recognition by the state as citizens. When the latter's own children register their deaths, they (although not their wealth) pass beyond the state's all-encompassing gaze. But how did this formal system for the ascription of citizenship come about?

Although the modern system of civil registration of births, marriages and deaths is a nineteenth century creation, the history of the registration of vital events in England and Wales can, in fact, be traced back to 1538. This was when Thomas Cromwell, acting as Henry VIII's vicar-general, sent a series of injunctions to bishops of the Church in England. These included specific instructions to parish priests on teaching the people the rudiments of the faith; preaching Scripture; guarding against superstitious ceremonies; the placing of an English Bible in each church; and on the keeping of registers of the baptisms, weddings and funerals at which the clergy officiated.[1] This was part of the process by which the medieval Church was brought under the control of the Tudor monarchs. The injunction respecting registration was repeated, and penalties for neglect prescribed, by an Act of 1597, which also directed that transcripts of the registers should be sent annually to a diocesan registrar.[2] Such decentralised arrangements for carrying out a

---

[1] Elton, *Policy and police*, p. 254.
[2] Nissel, *People count*, p. 6.

state policy were typical of governance in early modern England and Wales – the central government might lay down procedures but they were to be carried out by members of the local social elite.[3]

This parochial system remained the basis of registration until the passing of the Registration and Marriage Acts in 1836. The former Act laid down that the Poor Law unions, which had recently been established by the 1834 Poor Law Amendment Act, should be the basis of superintendent registrars' districts. These were to be further subdivided into registrars' districts for the purposes of registration. Births and deaths were to be registered by householders and next of kin with the local registrars in a prescribed format within set periods, after which a fee would be payable. The death certificate, for example, asked for the name, age, sex, and rank or profession, of the deceased, and the date and cause of their death. Clergy officiating at marriages were to keep a register of such events, and send copies of this to the local superintendent registrar on a quarterly basis. Local registrars were to keep copies of these certificates in their local registers. Duplicates of the entries in the registers were also to be sent to the central GRO in London, which was to have a Registrar General at its head. The latter was to be responsible for the oversight of the local registration system, although the registrars were to be appointed by the local Poor Law guardians. The 1836 Marriage Act, on the other hand, laid down that the calling of banns in Church was to be superseded by registrars' certificates. Places of worship were to be registered by the registrars, and marriages could be performed in such places in their presence, and were to 'be good and cognizable in like manner as marriages celebrated before the passing of this Act, according to the rites of the Church of England'. The practice of civil marriage before registrars was also established.

Together these two Acts ushered in a new epoch in the history of the registration of vital events in England and Wales. But what, one may ask, were the reasons for these sweeping changes?

---

[3] Higgs, *The information state in England,* pp. 28–44.

Civil registration: religious emancipation and the statistical sciences

Modern historical accounts tend to explain these developments in terms of either religious emancipation, or of the requirements of scientists for statistical data, or a combination of both. A survey of this historiography is of importance here, if only to point up its limitations.

In the eighteenth and early nineteenth centuries, Protestant Dissenters and Roman Catholics who failed to conform to the tenets and rites of the Church of England ('Nonconfromists'), had a number of grievances with respect to parochial registration. By law, and particularly under Sir George Rose's Parochial Registers Act of 1812, the church registers of such minorities were not admissible in court as evidence of births, marriages and deaths. Only those maintained by the clergy of the Church of England could be presented in court as legal documents. Additionally, under Hardwicke's Clandestine Marriages Act of 1753[4], except in the case of Jews and Quakers, legal marriages had to be carried out according to the rites of the Church of England. Dissenters who married using their own ceremonies alone were not, therefore, legally married, and many were forced to undertake a separate marriage ceremony before an Anglican clergyman.[5] This situation plainly annoyed Nonconformists, who had been in bitter conflict with the Established Church for centuries, and whose numbers and influence were growing in the period. It has been argued by scholars that this led to parliamentary legislation in the early nineteenth century to remove such grievances at the same time that other civil disabilities, such as the ban on Nonconformists attending the Universities of Oxford and Cambridge, were being lifted.[6]

---

[4] For the passage of the Act, see: Lemmings, 'Marriage and the law in the eighteenth century'.
[5] *Report of the select committee . . . on the general state of parochial registries . . .* , pp. 6, 75.
[6] This argument is rehearsed by Finer, *The life and times of Sir Edwin Chadwick*, p. 125; Lewis, *Edwin Chadwick and the Public Health Movement*, p. 30; Flinn, 'Introduction', pp. 27–8; Glass, *Numbering the people*, pp. 118–9; Cullen, 'The making of the Civil Registration Act'; Eyler, *Victorian social medicine*, p. 37–8; Nissel, *People count*, pp. 10–11; Goldman, 'Statistics and the society of science', p. 418.

Michael J. Cullen has produced the most sustained argument along these lines. He traces the genesis of the 1836 Acts to a Bill that Lord Nugent introduced into Parliament in 1832 to set up a general registry of births. Nugent pointed out, 'the legal problems affecting all because of the dubious status of the Dissenters' registers as evidence in courts of law.'[7] On the failure of this Bill, John Wilks, a radical MP, proposed the establishment of a Select Committee to inquire into the state of parochial registration. Wilks was leader of the Protestant Society for the Protection of Religious Liberty, and a key advocate of religious emancipation in Parliament. According to Cullen, this Committee, which reported in 1833, was essentially concerned with removing the religious disabilities laid on Nonconformists. The Committee concluded in its report that the system of registration via the Established Church was inequitable, and recommended the establishment of a national, civil system along the lines of that subsequently set up in 1836, although it did not identify who should run it.[8] Although the report referred to the need to improve the security of property through the better recording of lines of descent, Cullen saw this as merely a tactical device on Wilks' part to construct 'the semblance of a coalition' within the Committee.[9]

In 1834 a Registration Bill was introduced into Parliament by Lord Brougham based on the Select Committee's recommendations, and with the role of the registration officials being performed by the local tax gatherers. Cullen argues that when Brougham spoke to the Bill he, 'stripped away much of the national interest wrapping paper and based his case entirely on the desirability of remedying Dissenters' grievances . . .'.[10] As one might imagine, the suggestion that registration should be linked to tax gathering was unacceptable, and the Bill was withdrawn. However, it appears to have formed the basis of the Whig government's Registration Bill of 1836. Cullen notes that when

---

[7] Cullen, 'The making of the Civil Registration Act', pp. 42–3.
[8] *Report of the select committee . . . on the general state of parochial registries . . .* , pp. 9–11.
[9] Cullen, 'The making of the Civil Registration Act', p. 45
[10] Ibid., p. 50.

introducing this Bill into the Commons, the Home Secretary, Lord John Russell, stressed the need to remove the religious disabilities on Dissenters, and much of the subsequent debate revolved around this issue.[11] According to Lawrence Goldman, the passing of the Act provided, 'an answer to the long campaign of Dissenters that they should be allowed to register baptisms, marriages and burials through their own places of worship, and that their registers be accorded the same status in law as Anglican records.'[12]

But this, of course, was exactly what the 1836 legislation did not do. Instead the Registration Act created, at considerable expense, a national, centralised system for the secular registration of births and deaths. Indeed, it was the Act's opponents, including the Archbishop of Canterbury, who suggested the creation of a parallel system of legal Nonconformist registration via their own church records as an alternative to civil registration.[13] This measure of religious emancipation was partly achieved by a separate set of legislation resulting from the deliberations of the Commissioners 'into the state of non-parochial registers', who reported some years later.[14] They advocated the authentication by themselves of well-kept Nonconformist registers, and their removal to the GRO for safekeeping. The Commissioners' recommendations were carried into effect by the 1840 Non-Parochial Registers Act and the 1858 Births and Deaths Registration Act. The 1898 Marriage Act enabled Nonconformist ministers to act for the first time as 'authorised persons' for the registering of marriages in their own chapels. The creation of a civil registration system would, therefore, appear to be a somewhat surprising outcome for a purely religious campaign. Cullen also admits that Nonconformist members of Parliament had ceased to have much influence over

[11] *Hansard*, 3rd series, Vol. 31, Feb. 12, 1836, cols 367–80; *Hansard*, 3rd series, Vol. 32, 15 April, 1836, cols 1088–9; *Hansard*, 3rd series, Vol. 34, 6 June, 1836, cols 133–43; *Hansard*, 3rd series, Vol. 34, 28 June, 1836, cols 1011–21; *Hansard*, 3rd series, Vol. 35, 11 July, 1836, cols 79–89; *Hansard*, 3rd series, Vol. 35, 21 July, 1836, cols 375–6.
[12] Goldman, 'Statistics and the society of science', p. 418.
[13] *Hansard*, 3rd series, Vol. 34, 6 June, 1836, col. 133; *Hansard*, 3rd series, Vol. 35, 11 July, 1836, col. 83
[14] *Report of the commissioners . . . into the state of non-parochial registers.*

the legislation in the parliamentary session of 1836.[15] In fact, Cullen's emphasis on the purely religious aspects of the parliamentary events leading up to the passing of the Registration Act does not bear close scrutiny. As will be argued below, although Nonconformists took advantage of the establishment of the civil registration system for their own purposes, this was not the only, or indeed the main, reason for its creation.

Other historians have explained the origins of the civil registration system in terms of pressure from doctors, demographers and actuaries for accurate statistical data for their disciplines. According to John Eyler, in his magisterial intellectual biography of William Farr, 'From the perspective of the technical expert – physician, actuary, lawyer – the Registration Act marked the end of the first major campaign to remedy the long acknowledged deficiencies of existing English vital statistics'.[16] In addition, D.V. Glass points to the enthusiasm of the medical profession for the collection of cause of death data, and to the deficiencies in contemporary life tables for the calculation of commercial insurance policies.[17] The Select Committee on Parochial Registration called medical authorities to testify before it, who in its own words, 'have long desired ampler and more accurate information on the extent and causes of mortality . . .' . They also took evidence from Adolphe Quetelet, the great Belgian statist, on the statistical uses to which registration data could be put for insurance purposes.[18] Indeed, John Eyler describes Quetelet as the Committee's 'star witness'.[19] The Committee also published a resolution of the Provincial Medical and Surgical Society, as an appendix to their report, to the effect that, 'great benefit might be expected to accrue to medical science, and consequently to the community at large, if arrangements were made for recording the *causes of deaths* in the local registers of mortality.'[20]

---

[15] Cullen, 'The making of the Civil Registration Act', p. 53
[16] Eyler, *Victorian social medicine*, p. 37. See also Goldman, 'Statistics and the society of science', p. 415; Cullen, *The statistical movement in early Victorian Britain*, p. 29; and Schürer, 'The 1891 census and local population studies', p. 20.
[17] Glass, *Numbering the people* , pp. 126–7, 141–2.
[18] *Report of the select committee on parochial registries*, pp. 6, 121.
[19] Eyler, *Victorian social medicine*, p. 42.
[20] *Report of the select committee . . . on the general state of parochial registries . . .* , p. 174.

However, as will be shown below, these scientific concerns were of secondary importance in the Select Committee's deliberations. There was also very little discussion of such statistical matters in the parliamentary debates on the various Bills relating to registration. Similarly, cause of death data was not a feature of either Nugent's or Brougham's Bill, and was not included in the death certificate proposed in the original Registration Bill introduced into Parliament in February 1836.[21] As will be described in more detail in Chapter 2, the question on cause of death was only inserted into the death certificate in rather obscure circumstances during the course of the Bill's passage through Parliament.[22] Even then the information on cause of death did not have to be supported by a medical certificate until 1874. The 1836 Registration Act itself did not specifically authorise the production of statistical tables beyond stating rather lamely, 'That the Registrar General shall send once in every year to one of the principal secretaries of state a general abstract of the numbers of births, deaths, and marriages registered during the foregoing year, in such form as the said secretary from time to time shall require'.[23] The subsequent development of the GRO's statistical functions was, it will be argued below, mainly a function of later administrative action rather than a direct consequence of the legislative process.

Property rights and places of record

If neither religious movements or scientific curiosity appear to explain adequately either the events leading up to the establishment, or the specific form, of the 1836 civil registration system, then what did lie behind these developments? In order to

---

[21] PP 1831–2, I, pp. 265–72; PP 1834, III, pp. 459–477; PP 1836 I, f 309–26. This has been noted by Finer, *The life and times of Sir Edwin Chadwick* , p. 125; Lewis, *Edwin Chadwick and the Public Health Movement*, p. 31; Glass, *Numbering the people*, pp. 139–40; Eyler, *Victorian social medicine*, p. 45; Cullen, 'The making of the Civil Registration Act', pp. 55–8.

[22] The published volume of Parliamentary Papers has a Bill of 21 June with no cause of death (PP 1836, I, p. 366), and one of 3 August with cause of death (PP 1836, I, p. 388). However, an unpublished version of 21 July 1836 held in the House of Lords Record Office also includes cause of death.

[23] 6 & 7 Will IV c 86 s vi

answer this question it is necessary to place the whole question of registration in a broader context – that of the creation of institutional structures for the protection of property rights.[24]

Insecurity in the ownership and transfer of property was a general concern of early nineteenth-century English society. As the first report of the Commissioners 'appointed to inquire into the law of England respecting real property' noted in 1829:

> the modes by which estates and interests in real property are created, transferred and secured, are exceedingly defective, and require many important alterations.[25]

This feeling of insecurity, which appears to have applied to all forms of property, probably had a variety of causes – the fluidity of the land market[26]; the boom and bust of agriculture during and after the Napoleonic Wars[27]; the increasing complexity of the disposition of property via marriage settlements and wills[28]; and the legal uncertainties surrounding the new forms of property and financial instruments spawned by the Industrial Revolution.[29] The perceived deficiencies of the existing ecclesiastical system for registering baptisms, marriages and burials were added to these concerns. Since, as historical demographers have noted, the interval between births and baptisms was not stable, the date of the latter gave no reliable information on the date of natality, or of conception.[30] This was a serious problem under a system of entail, where property was bequeathed in wills to legitimate offspring, rather than to children conceived out of wedlock, or where property was to pass to the first born amongst a range of possible heirs. Similarly, under the custom of 'tenancy by courtesy' if a woman possessed of an estate by legal settlement died, the estate passed from her

---

[24] An earlier version of the following arguments can be found in Higgs, 'A cuckoo in the nest?'.
[25] *First report of the commissioners . . . appointed to inquire into the law of England respecting real property*, p. 7.
[26] Daunton, *Progress and poverty*, pp. 76–8.
[27] Thompson, *English landed society*, pp. 212–37.
[28] *Report of the select committee . . . on the general state of parochial registries . . .* , p. 69.
[29] Manchester, *Modern legal history*, pp. 1–6.
[30] Wrigley and Schofield, *The population history of England*, pp. 96–100.

husband to her heir-at-law, unless there had been a living child born legitimately to the couple, in which case the husband acquired a life-interest in the property. It did not matter if the child died within minutes of birth.[31] Knowledge of the exact dates of marriage, birth, and death were thus of crucial importance in a society to which the ownership of private property was so central. Since the transmission of property between generations was also of fundamental importance to the maintenance of the fabric of the middle-class family, uncertainty as to title threatened this most vital of nineteenth century social institutions.

There was also widespread discontent with the complexities, arcane trappings and dilatoriousness of the central courts of law, which Jeremy Bentham excoriated in so many of his writings[32], and Charles Dickens satirised in his portrait of the case of *Jarndyce v Jarndyce* in his novel *Bleak House* published in 1852–3.[33] Dickens had, of course, been an attorney's clerk and court reporter in the 1820s, and was fully conversant with the oddities of the English legal system. These concerns led to an extensive process of legal reform, associated with lawyers such as Lord Brougham, which in the course of the Victorian period saw such developments as the elaboration of the classical theories of contract and limited liability; new laws relating to trades unions, divorce and women's property rights; the amalgamation of the higher courts into the Supreme Court of Judicature; and the establishment of the county court system.[34] A widespread, complex movement to establish state institutions for the recording and preservation of titles to property was linked to this process of legal reform. The mid-Victorian period saw the establishment of numerous such repositories clustered around the inns of court and Supreme Court of Judicature in the

---

[31] Crowther and White, 'Medicine, property and the law', pp. 859–60.
[32] Postema, *Bentham and the common law tradition*.
[33] Dickens, *Bleak House*.
[34] Manchester, *Modern legal history, passim*; Holdsworth, *A history of English Law*, vol. 13, pp. 3–308; Stewart, *Henry Brougham*, pp. 179, 220, 233–8, 258–9, 277–85, 348–9. For a contemporary view of this process see John Stuart Mill in his essay on Bentham which appeared in the *London and Westminster Review* of August 1838; Mill and Bentham, *Utilitarianism and other essays*, pp. 158–61.

Strand, including the Public Record Office, the Patent Office, the Land Registry, and the Central Probate Registry.[35]

The passage of the 1836 Registration Act needs to be seen in the context of these developments. It should be noted at the outset that the original system of ecclesiastical registration was introduced in 1538 to underpin title to property during another period of great change in land ownership, the Dissolution of the Monasteries. In December of that year Thomas Cromwell issued a circular to justices of the peace denying rumours that ecclesiastical registration was for the purpose of imposing taxes. He stated that the true reason for its introduction was 'for the avoiding of sundry strifes, processes and contentions rising upon age, lineal descent, title of inheritance, legitimation of bastardy, and for knowledge whether any person is our subject or no.'[36] Similarly, the objection of Nonconformists to the system of ecclesiastical registration via the Church of England in the nineteenth century, was that they were being forced to go through religious ceremonies of which they did not approve in order to protect their rights in property by legally recording lines of descent.

The early nineteenth century saw several attempts to improve the protection of property rights via the registration of baptisms, marriages and burials. These included a Bill of May 1824 for the establishment of a Metropolitan Register Office, in order to remove the 'great expense and inconvenience in tracing pedigrees, and in investigating and completing titles to real and personal property ...'. By concentrating such records in the Metropolis, rather than leaving them scattered in diocesan record offices, it was hoped to make legal searches easier and cheaper.[37] This theme was taken up by the Real Property Commission, which issued four reports in the period 1829 to 1833.[38] In their first report the Commissioners advocated the

---

[35] Higgs, *The Information State in England*, pp. 80–1.

[36] Elton, *Policy and police*, pp. 259–60.

[37] Bill of 5 May 1824, 'To authorize the establishment of a Metropolitan Register Office for concentrating and preserving the registers of baptisms, marriages and burials in England': PP 1824, II, p. 672.

[38] *First report of the Commissioners appointed to inquire into the law of England respecting real property; Second report of the Commissioners appointed to inquire into the law of England*

establishment of a registry of wills and deeds, by which, 'the investigation of titles would be materially abridged and simplified, and fraud in all transactions respecting real property would be effectually prevented.'[39] By their second report they were talking of establishing a 'General Registry of Deeds and Instruments relating to Land', which would hold the records of the courts of law; of births, marriages and deaths; and of wills.[40] It was believed that the resulting improvement in the security of property would lead to an increase in its value.[41] This was similar to Jeremy Bentham's proposal for a 'General Register of Real Property', which he described in a letter to the Commissioners, and which they later published.[42]

It was in this context that Lord Nugent introduced his Bill in 1832 'to provide means for the general registration of births, and for a better and more direct evidence in all questions of birth, age and relationship, in time to come . . .'.[43] Nugent argued that the existing system of registering baptisms gave no direct legal proof of the date of an individual's birth, and proposed a system whereby parish clerks would keep registers of births and send copies to the High Court of Chancery. Although the problems associated with poor registration were most keenly felt by Nonconformists, especially Baptists who practised adult baptism, Nugent argued that they affected all classes in the community.[44] When John Wilks intervened in subsequent debates to propose the establishment of the Select Committee on Parochial Registration, this was a masterly exercise in jumping on the property-recording bandwagon. He noted that the existing system of registration undermined the property rights

---

respecting real property; *Third report of the Commissioners appointed to inquire into the law of England respecting real property*; *Fourth report of the Commissioners appointed to inquire into the law of England respecting real property*.

[39] *First report of the Commissioners appointed to inquire into the law of England respecting real property*, p. 60.
[40] *Second report of the Commissioners appointed to inquire into the law of England respecting real property*, passim.
[41] Ibid., p. 18.
[42] Bentham, 'Outline plan of a general register of real property'; *Third report of the Commissioners appointed to inquire into the law of England respecting real property*, pp. 430–50.
[43] PP 1831–2, I, p. 265.
[44] *Hansard*, 3rd series, Vol. 10, Feb. 23 1832, col. 686.

of Nonconformists, but that the deficiencies of the system affected, 'all persons possessing property, however humble or exalted, and to whatsoever religious denomination they belonged.' Cleverly he turned the particular problems of the Nonconformists into a general problem for all property owners by pointing out that even the most rigid adherents of the Church of England might buy property from Dissenters, and then find their right to that property challenged due to the defects in the title of the original owners.[45]

The work of the Select Committee on Parochial Registration was not, as Cullen argues, dominated by the need to placate the Nonconformists. The Committee did indeed find the existing system of registration by the Established Church 'exclusive and intolerant', but it also stressed that since it supplied:

> no adequate proof of pedigree, or means of proving or tracing ancestral descent . . . the value of property is diminished by the difficulties incident to its transfer, and the insecurity with which it is so often held and acquired.[46]

Its solution, civil registration with a central, metropolitan registry office, was plainly of a piece with previous suggestions for safeguarding titles to property. A detailed analysis of the Committee's workings also reveals the predominance of property issues. Thus, of the 129 pages of published evidence, some 43 pages, or a third, were provided by four legal witnesses, and a similar number of antiquaries and heralds produced another 12 pages. They testified to the failure of the existing parochial system to provide a cheap and reliable means for tracing the descent of property rights and titles. This hinged as much on the difficulty in any system of ecclesiastical registration of extrapolating back from baptisms to date of birth, as on the failure to capture vital events relating to Nonconformists.[47] At the same time 13 witnesses including clergymen, representatives of

---

[45] *Hansard,* 3rd series, Vol. 16, March 28 1833, cols 1209–21.
[46] *Report of the select committee . . . on the general state of parochial registries . . . ,* p. 9
[47] Ibid., p 76.

religious groups, parish clerks and archivists of ecclesiastical registries, provided 42 pages of testimony. On the other hand, an assortment of physicians, scientists, statisticians and actuaries could only muster 14 pages. One of the latter was Adolphe Quetelet but his testimony was of below average length, three pages compared to the average of 4.75, and much shorter than the average for legal witnesses at 10.6 pages. Quantity of testimony is not, of course, the same as importance but this hardly indicates that the Committee were eager to draw out the implications of Quetelet's evidence.

Moreover, much of the testimony of the religious witnesses related not to the unfairness of the existing system of ecclesiastical registration but to its general inefficiencies, the poor storage of registers, and the system's lack of comprehensiveness. The chaplain to the bishop of London, for example, was asked pointedly:

> Are you not aware that several trials have turned upon the evidence of baptisms and registrations, where the register has not been produced, that collateral evidence has not been considered legal, and that considerable property has not been recovered in consequence of the want of that being legal evidence?[48]

Similarly, it was put to Dr James Moore, vicar of St Pancras, that, 'Should you think it desirable, with a view to the security of property, and the necessity of tracing pedigrees and descent, that there should be a more authentic registry of births of all persons?'[49] The animosity between Nonconformists and the Established Church, as well as a general reluctance of the poor to approach any religious denomination[50], led inevitably to the conclusion that only a civil system of registration could provide a proper underpinning for property rights.

When Brougham introduced his own Registration Bill early in 1834, there were already no fewer than three other Bills

[48] Ibid., p. 43.
[49] Ibid., p. 20.
[50] Ibid., p. 123.

relating to the preservation of deeds, wills, and instruments concerning real property being considered by Parliament.[51] Brougham initially stressed the advantages to Dissenters of his proposals to establish a centralised system of civil registration for births, marriages and births.[52] But when opponents of his Bill asked why the Dissenters could not simply set up their own system of ecclesiastical registration to run parallel with that of the Church of England, Brougham indicated that his Bill:

> was intended not only for the relief of Dissenters, but of the whole community . . . At present there was no records of births and deaths, and the great object of this Bill was, to supply that defect, which was severely felt in cases of title and other cases involving property.[53]

When Brougham claimed that his Bill would provide relief to Nonconformists he was, it should be noted, referring almost exclusively to marriage. This was understandable since of the three events to be covered by the new system of civil registration – birth, death and marriage – only the latter was a religious rite. When the Whig government introduced its own Registration Bill into Parliament in February 1836, it took the precaution of also tabling a separate Marriage Bill in which the contentious issues respecting this crucial bone of contention between Nonconformists and Anglicans were addressed. Indeed, so separate were the two Bills that the registrar of marriages appointed under the 1836 Marriage Act was, in theory, a distinct officer to the registrar of births and deaths established under the 1836 Registration Act. Indeed, only just over half of the registrars of marriages appointed by 31 December 1838 were also registrars of births and deaths.[54]

Inevitably, the two Bills became inextricably linked in the ensuing debates in Parliament, especially since Lord John Russell, when introducing them together, stressed the need to

---

[51] PP 1834, III, pp. 563–652.
[52] *Hansard*, 3rd series, Vol. 23, May 13 1834, cols 940–2.
[53] *Hansard*, 3rd series, Vol. 24, July 2 1834, cols 1073–77.
[54] GRO, *1st annual report of the Registrar General* (hereafter *ARRG*), pp. 5–6.

meet the Nonconformists scruples with regard to marriage.[55] Most of the opponents of the Act did not object directly to the emancipation of the Nonconformists, however, but to the possible secularisation of marriage and name-giving the new system entailed. This it was feared, and perhaps with some justice, would lead to the general abandonment of links with any church.[56] The resulting confusion exasperated supporters of the Registration Bill such as Dr Stephen Lushington, who was forced to interject that:

> He utterly denied that the object of the [Registration] Bill was to afford relief to the Dissenters. He considered the question embraced in the Bill to be one of great national importance, and to those who understood anything about the difficulties which were experienced in the tracing of pedigrees its advantages would be too manifest to need explanation.[57]

Lushington was an active supporter of the removal of the disabilities on Dissenters but was also a judge in the Consistory Court of the Diocese of London that dealt in matrimonial and testamentary suits. As a Whig active in the reform of the legal system, the need to provide institutions for the proper recording of property rights was plainly uppermost in his mind.[58] Despite this and other explicit denials, however, the passage of the Registration Act came to be linked in the public mind with the relief of the disabilities on Nonconformists enshrined in the Marriage Act, and some historians have perpetuated this confusion.

The argument made here, that the 1836 Registration Act was essentially about property rights, is consonant with the views of many of the principal figures involved in the work of the registration system and the generation of medical statistics in the

---

[55] *Hansard*, 3rd series, Vol. 31, Feb. 12 1836, cols 372–7.
[56] *Hansard*, 3rd series, Vol. 34, 13 June, 1836, cols 490–4; *Hansard*, 3rd series, Vol. 34, 28 June, 1836, col. 1017.
[57] *Hansard*, 3rd series, Vol. 34, June 6 1836, cols 139–43.
[58] Waddams, *Law, politics and the Church of England*, pp. 14–21, 57–61, 160–93.

period under consideration in the present work. One might note, for example, the arguments of John Southerden Burn in his *The Marriage and Registration Acts with instructions, forms, and practical directions for the use of officiating ministers, superintendent registrars, registrars, etc.*, published in 1836. Burn had given evidence to the 1833 Select Committee on no fewer than three occasions regarding the law on ecclesiastical registration. He was indeed the only witness to be called more than once, and provided 11 per cent of the entire testimony. Burn saw the Registration Act in terms of property and family relationships:

> Such are the ramifications of society, such the reverses of fortune, that no person, however indigent, should be indifferent to the provisions of this Bill. It is not an infrequent occurrence for a mechanic, a labourer, a soldier, or a sailor, to leave his native town or native land, and, after amassing wealth, to die without a will, leaving his property to be divided amongst persons who, though but distantly related, may prove to be his next of kin.[59]

In this manner, the institutional underpinning of middle-class property rights was generalised as a basic human right of advantage to all social classes. On the other hand Burn believed that the Marriage Act was, 'a measure conceded to the several classes of Dissenters', which enabled them to marry at their own places of worship.[60]

Similarly, George Graham, the Registrar General from 1842 to 1879, declared before the Royal Sanitary Commission in 1869 that the 'great object' underlying the 1836 Act was the need to create a centralised and reliable means of protecting the property rights of citizens. This in turn, affected 'the interests of their kinsmen, their neighbours, and the community at large. For all the rights of succession to property depend on the legitimacy of birth, the validity of marriage, and proof of death . . .'.[61] Graham plainly saw the GRO's registration functions in terms of the

---

[59] Burn, *The Marriage and Registration Acts*, p 6.
[60] Ibid., p. 7.
[61] GRO, *31st ARRG for 1868*, pp. 290–1.

broader institutional movement to underpin property rights. For example, when referring in one of his *Annual reports* to the establishment of the Central Probate Registry in Knightrider Street in 1861 he claimed that the bringing together of probate and registration records on one site would greatly facilitate the transfer of property rights between generations, 'the one [set of records] containing proofs of death and heirship, and the other containing the authority for transferring property at the death of its possessors to their successors'.[62] This useful juxtaposition of the two institutions of record in Somerset House was achieved before his retirement at the end of the 1870s. Nor was he the only Registrar General to view the matter in this light. When Sir Bernard Mallet, head of the GRO from 1909 to 1920, gave his presidential address to the Royal Statistical Society in 1916, he noted that the development of the Office's scientific apparatus was not a necessary consequence of the passing of the 1836 Act. Indeed, 'The Registration Acts were primarily concerned with the record, the statistical side being ignored . . . The development, therefore, of the statistical work of the office has been due to administrative and not to legislative action.'[63]

Protecting the rights and obligations respecting property across the generations was more fundamental to the origins of the GRO and civil registration than the generation of statistics.

The statistical legacy of the 1836 Registration Act

The dependence of the GRO's future statistical functions upon a system for registering property rights, set it apart from many statistical bureaux on the Continent. The Statistique Générale de la France, which became a permanent body in 1833[64], and many similar European organisations, were not central repositories for nominal-level data. Rather they operated by sending out enquiries to local worthies and officials, and then collating the

[62] GRO, *22nd ARRG for 1859*, p. xlv.
[63] Bernard Mallet, 'The organisation of registration', pp. 4–5.
[64] Desrosieres, 'Official statistics and medicine in nineteenth-century France', pp. 515–18.

replies.[65] This meant that the continental statistical bureaux could cover a far greater range of subjects but lacked the solid database of information that underlay the GRO's work. Their activities were at once more spectacular, and perhaps less reliable. Moreover, by accepting the centrality of the recording of property rights in the establishment of the civil registration system in 1836, one can explain some of the features of the statistical data published by the GRO that have puzzled scholars. These include the under-registration of vital events due to the lack of compulsion in the registration system until 1874; the failure to register stillbirths until the twentieth century; and the lack of information on morbidity.[66] These 'deficiencies' only appear so if one assumes that the *raison d'etre* of the registration system was the generation of data for demographic and medical research.

There were certainly calls in Parliament in the 1830s for the establishment of a compulsory system of registration.[67] Lord John Russell, however, pinned his faith on the evident self-interest of citizens in safeguarding their family property rights to ensure compliance with the 1836 Act. As he explained in Parliament, he was 'quite sure that when the plan was established, the advantages attending it would be so obvious, and would soon be felt by all classes of persons, they would so soon perceive the benefit of having their children's' names inserted in the general register, that it would not be long before everyone would be willing to concur in carrying out the plan.'[68] This was, of course, a classic example of what his contemporary, Jeremy Bentham, described as the 'duty and interest junction principle' – the framing of laws in such a way that it was in the citizen's interests to observe conduct that it was his or her duty to obey.[69] That the protection of property rights mainly benefited

---

[65] Perrot and Woolf, *State and statistics in France*, pp. 156–8.
[66] Glass, *Numbering the people,* pp. 181–205; Eyler, *Victorian Social Medicine*, p. 45, 61; Hardy, 'Death is the end of all disease'; Luckin, 'Death and survival in the city', p. 55.
[67] *Hansard,* 3rd series, Vol. 10, Feb. 23 1832, col. 688; *Hansard,* 3rd series, Vol. 13, June 20 1832, cols 938–41.
[68] *Hansard,* 3rd series, Vol. 31, Feb. 12 1836, col. 371.
[69] Harrison, *Bentham,* pp. 117–34, 268–71.

the middle classes was obscured by the positing of that interest as a basic human right. Nor, as will be discussed below, was the introduction of compulsion under the 1874 Registration Act primarily a consequence of the requirement to provide improved data for research purposes. In fact, the crucial factor appears to have been the need to enforce legal obligations with respect to the vaccination system, and the protection of the rights to life of infants.

The omission of the recording of stillbirths is also understandable given that the 1836 registration system was designed to record the natality of *legal* persons. Although William Farr advocated the registration of stillbirths for statistical purposes in the *Annual report* published in 1866, George Graham, the Registrar General, blocked the proposal. As he explained to the Home Office the following year, he believed that to 'investigate every miscarriage and every abortion and the exact time of conception and the precise period of gestation appears to me an indelicate, indecent, nasty enquiry', and he feared that the inception of such a policy would undermine the willingness of parents to register births at all.[70] He was supported in this by William Ogle, the GRO's Superintendent of Statistics from 1880 to 1893, who testified before the 1893 Select Committee on Death Registration that it would be impossible to define or prove the period of gestation.[71] The Select Committee advocated registration, as had the Royal Sanitary Commission in 1871, but this procedure was only introduced in 1927.[72] The omission of stillbirths has, of course, significant implications for the calculation of infant mortality rates. Graham Mooney calculates, for example, that if one included estimated stillbirths in infant mortality rate in England and Wales in 1890 one would

---

[70] GRO, *27th ARRG for 1864*, p.191; Public Record Office (hereafter) PRO: HO 45/8044 1867 Registration of stillborn children.
[71] *First and second report of the select committee on death certification*, Q. 4088.
[72] Ibid., pp xxii–xxiv; *Second report of the Royal Sanitary Commission, Vol. 1*, p. 58; 16 & 17 Geo. 5, c.48. This was extremely late by European standards. Stillbirths were registered separately in Denmark, for example, from as early as 1800: Lokke, 'No difference without a cause', p. 81. The GRO did not begin to explore the concept of perinatal mortality, in terms of the combination of stillbirths and deaths in the first few weeks of life, until the 1950s: GRO, *Registrar General's Statistical Review for 1950: Medical*, p. 26.

raise the rate from 151 per 1,000 live births to 155. In the case of Norwich at the same date the rate would rise from 181 to 200.[73] Legal structures and social conventions plainly helped to determine the nature of the GRO's scientific output.

It is also difficult to see how a statistical apparatus that was parasitic upon the recording of births, marriages and deaths for the purposes of establishing lines of descent could be used to generate information on illness in addition to mortality. As will be discussed in the next chapter, at the outset of his career William Farr thought this was possible by assuming a fixed relationship between rates of morbidity and mortality from specific diseases. However, the system plainly could not gather information on non-fatal diseases, and the assumption of a fixed ratio of deaths to infections was not easily proved. Revealingly, although the 1871 Royal Sanitary Commission called for the registration of sickness, it was vague as to how this was to be achieved and what diseases should be recorded, and left its implementation to the GRO.[74] Graham had already testified to the Commission that he could not see how information on morbidity could be obtained via the registration machinery, and let the matter drop.[75]

In the absence of the generation of morbidity data via civil registration, other systems for this purpose had to be created. During the 1870s and 1880s, many towns in Britain introduced systems for the notification of infectious diseases via local Improvement Acts. Under the Infectious Diseases (Notification) Act of 1889 compulsory notification was introduced in London but was only optional elsewhere until the passage of a further Act in 1899.[76] The GRO began publishing figures for notifications of certain infectious diseases for London from 1895, a practice extended to the whole country in 1922. It was not until 1943, however, that the GRO, in conjunction with the Central Office of Information's Social Survey, began the regular and systematic surveying of illness and injuries. Similarly, it was only

---

[73] Mooney, 'Still-births and the measurement of urban infant mortality rates'.
[74] *Second report of the Royal Sanitary Commission, Vol. 1,* pp. 58–61.
[75] GRO, *31st ARRG for 1868,* pp. 288–90.
[76] Brand, *Doctors and the State,* pp. 117–8.

in 1948 that a single International List of the Causes of Death that could be used for both morbidity and mortality statistics was accepted by the World Health Organisation.[77] As will be noted below, this reflected the very different role played by the GRO in an age when health care had become a national right mediated via the central state.

If, however, statistical concerns played such a minor role in the establishment of the registration system, how did they come to gain a foothold in the early GRO? This will be the subject of the next chapter.

---

[77] Nissel, *People count*, p. 87; GRO, *Registrar General's Statistical Review for 1946–1947: Medical*, pp. 1–2, 278–89.

2

The genesis of state medical statistics

Edwin Chadwick, family obligations and the 1836 Registration Act

The idea that the central state gathers information from citizens in order to create statistical series is now taken for granted. This, however, was not the case in the late eighteenth and early nineteenth centuries. Rather than doing good, such exercises were usually assumed to be the prelude to exercises in taxation, military conscription, and state repression.[1] The establishment of a statistical function within the nascent GRO is, therefore, something that requires some explanation. This subject needs to be considered in two stages: first, how a question on cause of death was added to the death certificate, and secondly, how a Statistical Department came to be established within the GRO. Neither of these developments was a foregone conclusion when the Registration Bill was introduced into Parliament in 1836. Much the same could be said with respect to the expansion and proliferation of the GRO's statistical activities in the course of the nineteenth century, although this will be the subject of subsequent chapters. With hindsight History can seem a process of inevitable progress towards the present but on closer examination it becomes a muddle out of which a pattern emerges, in part, by chance.

As already noted, none of the proposed Bills for improving the registration of vital events introduced into Parliament in the 1830s, including the first version of what was to become the 1836 Registration Act, contained any reference to an enquiry respecting cause of death. It was only the second version of the

---

[1] Glass, *Numbering the people*, pp. 16–20.

latter Bill introduced in July 1836 that incorporated this into the form of the death certificate.[2] Various candidates could be put forward for the honour of instigating this innovation. The Provincial Medical and Surgical Association had, for example, written to the Select Committee on Parochial Registration in 1833 calling for the recording of cause of death.[3] Glass notes the claims made on behalf of Francis Bisset Hawkins, Professor of Materia Medica at King's College. There is also a letter in the Home Office papers dated March 1836 from a Henry Belinge advocating the introduction of cause of death registration.[4] But most scholars have accepted Edwin Chadwick's claim, in a letter to Dr John Laycock of 1844[5], that it was he who prevailed upon the parliamentary managers of the Bill to introduce the enquiry.[6]

As secretary to the Poor Law Commissioners, Chadwick was well placed to influence the contents of the 1836 Registration Bill. In its initial form, this envisaged the registration system as being part of the New Poor Law machinery. The Bill of February 1836 indicated that, 'the keeping of the said register [of births, marriages and deaths], and the control of the officers, clerks and servants employed about the same, shall be given to the Poor Law commissioners for England and Wales ...'. Section 4 of the original Bill lay down that the Crown would appoint the Registrar General but that all the other staff of the GRO should be appointed and controlled by the Poor Law Commissioners. The next section empowered either the Home Secretary or the Commissioners to make regulations for the management of the GRO, and for the duties of the Registrar General and his staff. Section 7 lay down that the local guardians

---

[2] House of Lords Records Office: Bill Papers.
[3] *Report of the select committee ... on the general state of parochial registries ...*, pp. Appendix 17.
[4] Glass, *Numbering the people,* p. 140; PRO: HO 44/29. Hawkins had published *The elements of medical statistics* in 1829.
[5] University College London Library: Papers of Sir Edwin Chadwick (1800–1890): 2181/3: Copybook III: letter to Dr Laycock, 13 April 1844, p 26.
[6] Finer, *The life and times of Sir Edwin Chadwick* , p. 125; Lewis, *Edwin Chadwick and the Public Health Movement* , p. 31; Glass, *Numbering the people* , pp. 139–40; Cullen, 'The making of the Civil Registration Act', pp. 55–8; Eyler, *Victorian social medicine,* p. 45.

of the Poor Law unions should appoint local relieving officers as local registrars and superintendent registrars, if 'properly qualified in the judgement of the [Poor Law] Commissioners or Registrar General, if there are no Commissioners'.[7] It was only in a revised version of the Bill, that of 18 April 1836, that the central supervision of the Commissioners was removed – the appointment of staff now being directly determined by the Registrar General and the Treasury, with the former making regulations for the workings of the registration system. The local guardians were still to appoint the local registrars, however, but now subject to the approval of the Registrar General.[8] It is difficult to give a satisfactory explanation for these changes since they do not appear to have generated any published debates in Parliament.[9] Perhaps there were fears that allowing the Commissioners to appoint thousands of local officials would give the Whigs excessive powers of patronage. But whatever the cause, the final arrangements plainly left the Registrar General in a position of unusual constitutional independence.

Cullen has argued that Chadwick's reasons for persuading Parliament in 1836 to insert a question on cause of death into the death certificate reflected his short-term, tactical requirements. Chadwick was in dispute with the medical profession over the terms on which medical officers would be appointed to the Poor Law unions, and, according to Cullen, 'By including the cause of death Chadwick could win the favour of the Provincial [Medical and Surgical] Association, which had wanted its insertion, as well as to provide an excuse for bribing

---

[7] A bill for registering births, deaths and marriages in England, 17 February 1836 PP 1836 I, f309–26.

[8] A bill for registering births, deaths and marriages in England, 18 April 1836 PP 1836 I, f329–47.

[9] Cullen discusses the negotiations between Chadwick and Lord Ellenborough over the appointment of local officers but says nothing about the role of the Commissioners: Cullen, 'The making of the Civil Registration Act of 1836', pp. 56–7. Chadwick plainly deprecated the removal of the Commission's central supervision: University College London Library: Papers of Sir Edwin Chadwick (1800–1890): 1733: Correspondence with Lord John Russell: letter of 3 July 1836. For the manner in which sloppy parliamentary drafting in the early nineteenth century could leave civil service bodies without adequate official oversight, see, Clark, 'Statesmen in disguise' pp. 29–30.

the medical officers with the offer of the registrarship.'[10] Eyler repeats Cullen's belief that Chadwick had cause of death inserted to placate the medical profession, but wisely adds, 'It is also important to remember that Chadwick's interest in the registration of death and its causes quickly outran these small political ends, as he realised the potential value such information might have for Poor Law administration and especially for sanitary reform.'[11] This would certainly fit in with what we know of Chadwick's interest in public health, based on his acquaintance with French literature on 'hygiène publique' in the 1830s.[12] Chadwick was also well aware that illness and death cost money in terms of charges on the poor rates.[13]

But if we see the new registration system as originally envisaged as part of the machinery for underpinning the Poor Laws, an additional reason for Chadwick's interest might be discernible. Thus, in a memorandum in the Home Office papers for 1835, the Poor Law Commission revealed that it was considering using local Poor Law medical officers to register deaths. The memorandum claimed that, 'the correct statement of the nature of the disease . . . would be of great importance for scientific purposes.' However, the 'scientific purposes' noted in the memorandum were not medical but 'the formation of tables of insurance'.[14] This may have been linked to a desire to facilitate the activities of working-class friendly societies in which the poor insured themselves against the loss of income occasioned by illness or death. These had long been seen by politicians as an important alternative to the Poor Law as a means of preventing destitution. By combining relief of poverty with an education in personal responsibilities and

---

[10] Cullen, 'The making of the Civil Registration Act', p. 58. In this he is followed by Brundage, *England's 'Prussian Minister'*. p. 79
[11] Eyler, *Victorian social medicine*, p. 45.
[12] La Berge, 'Edwin Chadwick and the French connection', pp. 23–41.
[13] The Library, University College London: Papers of Sir Edwin Chadwick (1800–1890): 69: undated paper entitled, 'On the expenses of excessive death rates'; Flinn, 'Introduction', p. 43.
[14] PRO: HO 73/51: Permanent Officers. The same document also implied that Poor Law officers might use the registration system for establishing settlement.

obligations, friendly societies were regarded as a means of both social and moral improvement.[15]

Thus, Sir George Rose had sponsored an Act in 1793 in order to place friendly societies on a better legal foundation, with the 'benevolent intention of improving the conditions of the working classes and with the prudent intent of reducing Poor Law expenditures.'[16] Between 1793 and 1819 there had been as many as six Acts of Parliament that had widened the regulation and legal rights of friendly societies. Under the 1819 Friendly Societies Act, the tables of deaths and morbidity used for actuarial calculations had to be approved by at least two actuaries or 'persons skilled in calculation.' At the same time, a number of select committees on friendly societies and life annuities reported in 1825, 1827, and 1829.[17] Finally, by the 1829 Friendly Societies Act official supervision of these self-help organisations was shifted to central government, and regular returns of actuarial tables were required from them. It is perhaps besides the point whether such measures could have solved the problems of poverty in practice. As William Cobbett exclaimed with respect to their advocates:

> What wretched drivellers they must be: to think that they should be able to make the pauper keep the pauper; to think that they shall be able to make the man that is half-starved lay by part of his loaf![18]

What is relevant here is that many middle-class reformers believed that friendly societies might serve this purpose, and were willing to establish institutional structures to facilitate working-class saving.

Chadwick's activities in the 1830s in connection with registration can be seen in these terms. His earliest published work

---

[15] For the emphasis placed on insurance and self-help in debates on welfare at this period, see Finlayson, *Citizens, state, and social welfare*, pp. 25–91; and Supple, 'Legislation and virtue'.
[16] Cowherd, *Political economists and the English Poor Laws*, pp. 8–9
[17] *Report of the Select Committee on Friendly Societies*; *Second report of the Select Committee on Friendly Societies*; *Report of the Select Committee on Life Annuities*.
[18] Cobbett, *Rural rides*, pp. 472–3.

was entitled *An essay on the means of insurance against the casualties of sickness, decrepitude, and mortality*, which had appeared in the *Westminster Review* of 1828. In this he praised working-class friendly societies for helping to improve the conditions of the poor, and called on the state to obtain 'all the data desirable to be deduced from the past experience of the casualties of sickness and mortality', in order to improve the workings of such societies. Unless friendly societies calculated their subscriptions and benefits using accurate life-tables, they were prone to financial collapse. Chadwick had his article reprinted in 1836 to coincide with the passing of the Registration Act, thus emphasising the latter's actuarial potential.[19] To this he appended a list of benefits that might flow from statistical data collected by the new system, amongst which advances in medical science were but one.

In addition to their utilisation in actuarial calculations relating to life assurance, cause of death data could be seen as a basis for the study of rates of illness and thus of sickness insurance. This rested on the contemporary belief, voiced both by the young William Farr and by John Finlayson, the chief actuary at the National Debt Office, that there was a fixed relationship between rates of morbidity and mortality from diseases.[20] As late as 1875 Farr could still claim that, 'There is a relationship betwixt death and sickness; and to every death from every cause there is an average number of attacks of sickness, and a specific number of persons incapacitated for work. Death is the extinction of pain'.[21]

Chadwick was, of course, a follower of Jeremy Bentham, having acted in his youth as the utilitarian philosopher's amanuensis. In his actuarial proposals Chadwick was drawing on Bentham's *Pauper management improved* of 1798, in which the latter had drawn a distinction between the sturdy beggar and the 'self-maintaining poor'. The former were to be controlled by

[19] Chadwick, *An essay on the mean of insurance*.
[20] Eyler, *Victorian social medicine*, pp. 108–9; *Report of the Select Committee on Friendly Societies* , pp. 416, 456–70.
[21] GRO, *Supplement to the 35th ARRG*, p. iv. For a discussion of the problems inherent in the extrapolation of morbidity rates from mortality rates see: Riley, 'The morbidity of medical practitioners'.

workhouses, in which conditions were to be worse, or 'less eligible', than those outside. The 'self-maintaining poor', on the other hand, were to be helped by friendly societies and 'Frugality Banks'. Bentham added that the data upon which actuarial decisions were being made were imperfect, 'nor is the deficiency such as can be supplied without the aid of government.'[22] Bentham may, in turn, have taken the idea from Condorcet, the French *philosophe*, who had raised the possibility of the poor insuring themselves against distress in his *Esquisse d'un tableau historique des progrès de'l espirit humain* published in 1795.[23]

Understanding the GRO's intended work as, in part, an institutional support for the friendly-society movement, as well as other forms of self-help, would help to explain some of Chadwick's other interventions in its development. Thus Chadwick urged Lord John Russell to appoint someone with actuarial skills as Registrar General[24], and put forward his friend Charles Babbage, the mathematician and computer pioneer, as a candidate.[25] Babbage was interested in insurance, and had been asked in 1824 to become an actuary and manager of a projected new life insurance company – 'The Protector'. Although nothing came of this, Babbage had subsequently written *A comparative view of the various institutions for the assurance of lives*.[26] Chadwick's interest in the GRO as an actuarial institution would also explain why the passage of 1844 in which Chadwick claimed responsibility for the insertion of cause of death into the 1836 Act was in a letter dealing with the work of actuaries.

The model of citizenship and of the central state being elaborated here was, of course, a very limited one. The state was not

---

[22] Bentham, *Pauper management improved*, pp. 180–1. William Farr took much the same stance in his Letter in the GRO's *12th ARRG*, pp. xxxvi–xxxvii. For the relationship between the coercive and supportive aspects of Chadwick's social policy, see: Donajgrodski, "Social police' and the bureaucratic elite', pp. 63–73.

[23] Condorcet, *Sketch for a historical picture of the progress of the human mind*, pp. 180–2.

[24] The Library, University College London: Papers of Sir Edwin Chadwick (1800–1890): 1733: Correspondence with Lord John Russell: letter of Sunday July 3 1836, p 17.

[25] Eyler, *Victorian social medicine*, p. 46.

[26] Hyman, *Science and reform*, p. 110; Campbell-Kelly, 'Charles Babbage and the assurance of lives'.

to concern itself with taking action but only with encouraging people to act as rational, independent agents within the market economy in support of themselves and their families. The role of government was to give citizens information in order to allow them to choose their place of abode, occupations, and investments, wisely. As Chadwick expressed this in his *Essay on the means of insurance*:

> The misdirection of these sympathies (of the middle classes for the poor), and their operation in inconsiderate charities and the profuse expenditure of the poor's-rates, have formed the most potent means of retarding the improvement of the labouring population; and it seems to us that the wealthy have yet to learn what are the means by which they may render the best services which means, we conceive, will be found to be, an acting with the labouring classes rather than for them, in enabling them to act for themselves, by provident institutions, securely based on sound knowledge of the nature of that of which we have treated.[27]

Data gathering was seen in terms of facilitating independent activity, rather than as overt control or coercion. But this emphasis on thrift as the means of working-class emancipation was one that appears to have been shared by the members of friendly societies themselves.

The creation of the GRO's Statistical Department

Such actuarial considerations were also to play a crucial role in the establishment of a Statistical Department within the GRO. There is little evidence, however, that such a division was originally seen as part of the Office's core structure. Thus, when the first Registrar General, Thomas Lister, wrote to the Treasury on 6 November 1837 to discuss the duties of the GRO, he omitted any mention of a statistical function. Rather, the business of the Office was to be divided into three branches – Correspondence;

---

[27] Chadwick, *An essay on the means of insurance*, p. 61.

Accounts; and that dedicated to the 'Arrangements & indexing of certified copies of all registers – including the admission of the public to search.'[28] Lister put in a revised bid to the Treasury in March 1838 but again made no mention of the need for a statistical unit.[29] Indeed, it was not until 19 May 1838, nineteen months after the GRO had started work, that Lister approached the Treasury for permission to employ someone for the 'difficult and important duty' of drawing up tables of the causes of death. He envisaged the potential post-holder as, 'belonging to the medical profession, or at least having a competent knowledge of diseases, and having given some attention to medical statistics.' Although Lister felt that the work would have to be undertaken each year, he did not foresee that it would, 'fully occupy the time of the person to whom it may be assigned.' As a consequence, he did not want to employ a permanent member of staff but only to pay someone 'in the first instance for only a time to be remunerated for the services specially required . . .'.[30] It should be noted that the appointee was not necessarily intended to produce tables showing variations in local mortality experience.

A few weeks later, in a letter of 4 June 1838, Lister turned his mind to the actuarial needs of the new Office, and explained to the Treasury that:

> In the abstract of all registered births, deaths and marriages which it will be my duty to prepare annually to be laid before Parliament it will be obviously important with reference to annuities and life insurance to exhibit the numbers of deaths which occur at every age in England and Wales, and in various parts of the same, showing the comparative mortality in towns and in the country, and in various agricultural, manufacturing and mining districts.

---

[28] PRO: RG 29/1, p. 16. 'Correspondence' meant the production of circulars, letters, etc., relating to the work of the registration system; 'Accounts' related to the payment of fees, expenses, salaries; 'Arrangements & indexing of certified copies' involved checking incoming copies of certificates of births, marriages and deaths, transcribing entries, cutting them up and arranging them alphabetically, and then copying them into index books.
[29] Ibid., pp. 26–9.
[30] Ibid., p. 35.

The genesis of state medical statistics

Again, he was not sure if this work would require full-time staff but requested funds to employ, 'two persons on whose accuracy entire reliance can be placed and who shall be bound to perform the whole of such duty before the month of December next ensuing.'[31] The potential for comparative statistical work on local mortality was seen at the outset, therefore, as part of the GRO's actuarial functions. It should also be noted that the production of actuarial tables was seen as requiring more staff resources than medical statistics, and that the staff to be employed on the former were not at first subordinated to a medical specialist. Lister's initial statistical project appears very similar to that of Chadwick in his *Essay on the means of insurance*.

The Treasury replied to Lister's second letter within a week of its receipt, and in an extremely positive manner. They agreed to fund the necessary staff for actuarial purposes because they considered 'it important that the information should be afforded . . .'.[32] It was another fortnight, however, and nearly six weeks after its original receipt, before the Treasury replied to Lister's communication on the subject of medical statistics. In this case the Treasury expressed itself very doubtful about making an appointment at all, on the not insubstantial grounds, 'that unless the information proposed to be given can be afforded with the certainty of entire accuracy, it would be better that it should be omitted, as being otherwise calculated to mislead in a matter of very great importance.'[33] Such a Treasury 'demur' did not mean a total rejection of the request but hardly indicates that medical statistics were seen as central to the GRO's work.[34] Lister wrote back on 30 June 1838 explaining that he had already persuaded the Presidents of the Royal Colleges of Physicians and Surgeons, and the Master of the Society of Apothecaries, to send medical practitioners a plea to provide accurate data on the death certificates.[35] The Treasury

---

[31] Ibid., p. 36.
[32] PRO: RG 29/5, p. 56.
[33] PRO: RG 29/5, p. 59.
[34] On the Treasury 'demur', see Wright, *Treasury control of the Civil Service*, p. 151.
[35] PRO: RG 29/1, p. 39.

gave way, 'under the circumstances', and sanctioned a salary of £150 to £200 to hire someone with medical knowledge.[36]

In this rather inauspicious manner, a vacancy for a medical statistician was created within the GRO. Into this opening stepped William Farr, who was to radically alter the Office's role and structure by his activities. Farr had been born in Shropshire in 1807, the son of a farm labourer, and received an education via the good offices of the local squire. He studied medicine in Paris under Pierre Louis, the French physician who first used mathematical methods in the quantitative analysis of diseases, and returned to England in 1830. Farr received the licentiate of the Society of Apothecaries, the lowliest medical qualification, in 1832, and set up in medical practice in London in 1833. Times were plainly hard for a poor doctor struggling to bring up a young family, and he increasingly turned his hand to medical journalism. Via his writings Farr contributed to the campaign for the reform of medical training and professional status that was such a feature of the period.[37] Increasingly, however, he was drawn to quantification, and in 1837 he contributed a section on vital statistics to John Ramsay McCulloch's *Statistical account of the British Empire*. In this he drew on material from both the actuarial and medical sciences.[38]

There is some debate over who recommended Farr to Lister for the post at the GRO. Noel Humphreys, who worked with Farr, claimed that this role was played by the physician, Sir James Clarke.[39] However, Edwin Chadwick also claimed the honour, and he was certainly in correspondence with Lister in 1836 over the possibility of appointing an unnamed medic to handle the data on cause of death.[40] Whatever the exact route by which Farr entered the public service, it should be noted that it was on

---

[36] PRO: RG 29/5, pp. 59–60.
[37] Eyler, *Victorian social medicine*, pp. 1–12.
[38] McCulloch, *A statistical account of the British Empire*, pp. 567–601.
[39] Humphreys, *Vital statistics*, p. xii. The claim is repeated by Flinn: 'Introduction', p. 27.
[40] The Library, University College London: Papers of Sir Edwin Chadwick (1800–1890): 2181/3: Copybook III: letter to Dr Laycock, 13 April 1844, p. 27; ibid., 1240: Correspondence with Thomas Henry Lister: letter of 18 October 1836.

the basis of a singleton post with no management duties. It was not until July 1839 that Farr was formally appointed to the permanent post of 'examiner and compiler of abstracts', with a complement of three clerks under him.[41] Only at this date, some three years after its inception, could a distinct and permanent Statistical Department be said to have existed in the GRO. In addition, the Statistical Department was not solely dedicated to medical work — the needs of actuaries for reliable lifetables also bulked large.[42]

Moreover, one should not over-estimate the resources of Farr's unit when first established within the GRO. In 1840 the Statistical Department was the smallest section of the Office, employing four clerks compared to five in Accounts, ten in Correspondence and no fewer than 29 in the Records Department.[43] Nor were Farr's clerks medical specialists. In July 1840 Lister asked the Treasury to confirm H. J. Edwards, who had received a medical education, as assistant to Farr. When the Treasury demanded on what authority Edwards had been appointed in the first place, Lister had to reply, rather shamefaced, that during his employment Edwards had not 'been acting under any promise or understanding that he would receive remuneration — that his services were tendered without previous solicitation . . .'. The Treasury subsequently turned down his appointment as 'very irregular'.[44] It was only during the course of the next decade that Farr received a modicum of medically qualified support within the Office.[45]

In mid-1839, when Lister published the first *Annual report of the Registrar General,* it contained sixteen appendices, only one of which was a 'Letter to the Registrar General' from William Farr on registered causes of death.[46] At this date it would have been difficult to imagine that Farr's Letter would grow in extent and importance so as to eventually form the body of the *Annual*

---

[41] PRO: RG 29/1, p. 61; RG 29/5, p. 95.
[42] For the role the medical profession played in the business of life assurance in the nineteenth century see, Dupree, 'Other than healing'.
[43] HMSO, *Imperial calendar for 1840*, p. 181.
[44] PRO: RG 29/1, pp. 85, 187; RG 29/5, pp. 128, 131.
[45] PRO: RG 29/1, pp. 278–9; RG 29/5, p. 296.
[46] GRO, *1st ARRG for the year ending 30 June 1838*, pp. 65–125.

report in its own right. This reflected the growth of the size and functions of the Statistical Department in the decades after its establishment, as will be described in subsequent chapters.

The decline of the GRO's insurance project

Given the centrality of actuarial considerations in the origins of statistical work within the GRO, the absence of the subject from some accounts of the GRO's origins and development appears inexplicable. In contrast to the Office's medical role, however, the story of its actuarial work is one of relative failure, and gradual eclipse. This helps, no doubt, to explain the omission.

The early *Annual reports,* however, bear testimony to the importance the Office attached to its actuarial functions, and to the role it played in initiating the GRO's examination of the local mortality experience. In his first *Annual report* Lister claimed that:

> In the Abstract of Deaths I have entered into more minute details exhibiting enumerations of deaths of persons of each sex at every successive year of age. Such details are of acknowledged value, as data for determining the laws of mortality – as bases for calculations materially affecting the interests of millions. Tables exhibiting the proportion of deaths at every successive year of age are among the most important materials from which are deduced the true principles on which should be founded the systems of life annuities and of life insurance and the rules of friendly societies established for the use of the poorer classes.[47]

To underline the importance of this activity he went on to quote from the Report of the Select Committee on Friendly Societies of 1827, and from John Finlayson's testimony before the Committee on Parochial Registration in 1833. Similarly, his successor George Graham declared in his first *Annual report*, with reference to the publication of ages at death in particular

---

[47] Ibid., p. 15.

districts, that friendly societies would be able to obtain from the local tables the means of adjusting their premiums according to the prevailing rates of sickness and mortality. These he noted could be a 25 to 50 per cent higher in some districts than in others, 'so as to affect materially the money value of assurance and the stability and prosperity of the societies'.[48] Much of Graham's own section in his second *Annual report* was given over to a discussion of the construction of life tables for actuarial purposes, and contained a simple 'English Life Table No. 1'.[49] Reporting of mortality by age and place was often more extensive than that by cause in the 1840s. In the *Annual report* for 1845, for example, there were over 100 pages of tables for the former, but only 16 pages by causes of death.[50]

These themes continued to run through the *Annual reports* of the 1840s and early 1850s. Thus, the *Annual report* for 1842 contained no fewer than 149 pages on actuarial aids and the calculation of insurance premiums.[51] Similarly, Farr's Letter in the *Annual report* for 1845 was mostly given over to a consideration of problems with the construction of Richard Price's Northampton Life Table, upon which much insurance work was based.[52] The Letter in the *Annual report* for 1849 was again concerned mainly with life insurance, including the publication of a second English Life Tables for males as a prelude to 152 pages of actuarial tables and calculations.[53] Two years later, Graham's section in the *Annual Report* for 1851, published in 1855, was mainly concerned with tables of occupational mortality, which were expressly seen as a an aid to the construction of occupational life tables.[54] In the same year a Treasury committee of inquiry into the GRO concluded that:

> The primary aim of this Department is to record all the births, deaths and marriages in England and Wales, without

[48] GRO, *4th ARRG for the year ending 30 June 1841*, p. 3.
[49] GRO, *5th ARRG for 1841*, pp. 16–49.
[50] GRO, *8th ARRG for 1845*, pp. 82–97, 166–268.
[51] GRO, *6th ARRG for 1842*, pp. 517–666.
[52] GRO, *8th ARRG for 1845*, pp. 277–356.
[53] GRO, *12th ARRG for 1849*, pp. 1–152.
[54] GRO, *14th ARRG for 1851* (London, 1855), p. xxiii.

distinction of class or religious persuasion, in order to furnish the means of tracing the descent of property, of calculating the expectation of life and the laws of mortality, and of ascertaining the state of disease and the operation of moral and physical causes on the health of the people and the progress of population.[55]

The ordering here of the functions of the GRO probably reflected the official estimate of their relative importance at this date.

The need to ensure that the information on mortality by occupations was consistent with the occupational data in the census also led Graham in September 1854 to ask the Treasury for financial support for the compilation of a dictionary of occupations. This was to enable clerks to place occupations under the correct headings in the occupational classifications being developed by the Office for census and registration work. Graham added that:

> without this assistance I cannot make as good use as I could wish of the facts which are in my possession as to the varying mortality in different professions, information which is desired by the public now that attention is much called to the benefits to be derived by the working classes from life insurance, and which should be given on the same principle as the numbers living in each profession as recorded in the census.[56]

Thus, the motivation for the GRO's systematisation of occupational classification systems, which was to play such an important role in the future analysis of British society, was actuarial rather than medical, economic or sociological.[57]

---

[55] PRO: RG 29/5, insert p. 385, p. 1.
[56] PRO: RG 29/1, pp. 448–9.
[57] There is no evidence for a formal occupational dictionary being used in the analysis of the 1851 census. A printed version of the 1861 dictionary survives in the PRO in London. The occupational classifications in 1851 must, therefore, have been fairly rough and ready. The work done on the dictionary presumably lay behind Farr's publication in the *1861 Census report* of an appendix entitled, 'The new classification of the people according to their employments': *1861 Census report*, pp. 225–48.

After the mid-1850s the pace of the Office's publication of actuarial material slowed noticeably, although it's efforts in the field were crowned in 1864 by the appearance of the *English Life Tables* as a separate volume.[58] The *Annual reports*, however, were now singularly bereft of direct actuarial references. That for 1865 contained extracts from an English Life Table No. 3[59], whilst a 'Healthy District Life Tables', with tables of annuities and premiums appeared in the *Annual report* for 1870. But the latter had been constructed as long ago as 1859 from observations extending over the period 1849 to 1853.[60] Life tables continued to appear in the GRO's *Decennial supplements* but these were expressly designed to reveal the increase in life expectancies in differing age groups, and thus the effects of public health reform, rather than to facilitate the calculation of insurance premiums.[61] In addition, the extent of the tables dedicated to the number of deaths by place contracted between the first and second of the two *Annual reports* published in 1855. The first, covering the year 1851, contained 60 pages of such tables, whilst the second, that for 1852, only four.[62] The latter became the pattern for subsequent *Reports* until the tables were dropped completely in 1883.[63] In the *Annual report* published in 1888, the third Registrar General, Brydges Henniker, could look back over the first 50 years of registration without even mentioning the GRO's actuarial functions.[64]

This did not mean, of course, that William Farr ceased to provide ad hoc actuarial services in a personal capacity to government departments and external bodies.[65] Indeed, Farr's Treasury superannuation file is mainly concerned with his work as an actuary. The file contains no fewer than three copies of an entry on Farr from the *Insurance Cyclopaedia*, in which references

---

[58] GRO, *English life tables*.
[59] GRO, *28th ARRG for 1865*, pp. lxvii–xci.
[60] GRO, *33rd ARRG for 1870*, pp. 441–57.
[61] GRO, *Supplement to the 35th ARRG*, p. ix.
[62] GRO, *14th ARRG for 1851*, pp. 28–82; GRO, *15th ARRG for 1852*, pp. 96–99.
[63] GRO, *44th ARRG for 1881*.
[64] GRO, *50th ARRG for 1887*, pp. xviii–xx.
[65] Eyler, *Victorian social medicine*, pp. 83–90.

to Farr's actuarial, life insurance and census work are underlined in red, apparently by his friend Florence Nightingale. But such material was collected to show his non-official activities in order to justify increased pension rights.[66] One should not confuse Farr's personal activities with the official work of the GRO as a whole. This was made very plain in November 1874 when the General Post Office wrote to the Treasury asking if Farr could act as an actuary in an arbitration case between it and the Great Eastern Railway. The Treasury wrote to Graham asking ominously, 'whether my Lords are to understand that Dr Farr is an officer who gives his whole time to the public service in your department.'[67] Graham had to write back defending Farr's record and indicating that 'under his instruction and guidance his assistants in the Statistical Branch of this office have obtained such proficiency, that his constant personal daily superintendence is not absolutely required.'[68]

The reasons for the gradual withering away of official insurance work within the GRO are plainly complex. In part this reflected the changing focus of interest within the GRO. Farr always appears to have seen his insurance work as secondary to the generation of medical statistics[69] but George Graham also came gradually to the same conclusion. Thus, under the impact of the use made of the data by the 1843 Royal Commission on the Health of Towns and Populous Places and in the 1848 Public Health Act, the Registrar General had come by 1850 to see the publication of mortality by place largely in terms of health statistics.[70] By the time of the *Annual report* for 1853, compiled in 1855, Graham was following Farr in comparing death rates in local areas with a Healthy Districts' norm of 17 per thousand in an attempt to measure insalubrity.[71] As will be discussed below, the very success of the GRO's medical propaganda was pulling its statistical work in this direction.

---

[66] PRO: PRO: T 1/12682/3906.
[67] PRO: RG 29/6, pp. 127–8.
[68] PRO: RG 29/2, p. 200.
[69] GRO, *8th ARRG for 1845*, p. 27.
[70] GRO, *10th ARRG for 1847*, p. ix.
[71] GRO, *16th ARRG for 1853*, pp. xv–xvi.

On the other hand, it is plain that the GRO's actuarial endeavours were not a great success in practical terms. As Eyler has noted, life tables for the whole population were not the sort of actuarial aids sought by commercial insurance firms in the early Victorian period.[72] On the whole, they insured the lives of the middle classes rather than of the populace in general, and could generate their own actuarial data.[73] Tables drawn up by the Institute of Actuaries from the experience of twenty life assurance offices were published in 1869 and commanded widespread confidence thereafter.[74] Similarly, friendly societies could not use the GRO's death data to work back to rates of sickness because, despite Farr's protestations, the relationship between rates of mortality and morbidity were not necessarily constant over time. Whilst adult mortality rates declined in the late nineteenth century, rates of morbidity may have increased in terms of duration of ill-health.[75] Nor did the GRO's data cover the numerous illnesses which contributed to unemployment, and thus to the levels of benefit payments, but which did not normally lead to death. In addition, many benefit societies were not locally based, but served a single trade in which its members 'tramped' over a wide area.[76]

The friendly societies also had an alternative source of actuarial guidance in the Registry of Friendly Societies (RFS). Under the 1829 Friendly Societies Act the rules of societies were to be confirmed by a barrister appointed to the National Debt Office (NDO). Under the provisions of the 1846 Friendly Societies Act, a separate RFS was set up to oversee the work of the friendly societies and to collect the actuarial data demanded from them by the Act. The first Registrar of Friendly Societies was John Tidd Pratt, who had been the officer responsible for friendly societies at the NDO.[77] Pratt did not look to the GRO

[72] Eyler, *Victorian social medicine*, p. 84.
[73] Pearson, 'Thrift or dissipation?'.
[74] Dupree, 'Other than healing', p. 87.
[75] Riley, 'Disease without death'; Riley, 'Ill health during the English mortality decline', p. 564; Riley, *Sick, not dead*. But see also: Woods, "Sickness is a baffling matter'.
[76] Hobsbawm, 'The tramping artisan', p. 313.
[77] Gosden, *The friendly societies in England 1815–75*, pp. 104–5.

to provide actuarial guidance but to his former colleague, Alexander Finlayson. Alexander was the son of John Finlayson, and had succeeded his father as actuary to the NDO in 1851. He published a series of analyses of the data received on individual society members[78] by the RFS, showing morbidity and mortality at particular ages nationally; by region; by settlement type (city, town, rural); by nature of work (light as opposed to heavy labour); and by specific trades. These also contained tables showing recommended weekly payments and returns on annuities.[79] It was these, rather than the GRO's life tables, that Tidd Pratt recommended to friendly societies as actuarial aids.[80] Eclipsed by medical statistics, and too theoretical for practical insurance purposes, the GRO's actuarial work faded away. That the GRO's Statistical Department did not fade away with it was due to the development of its work on medical and sanitary statistics, which will be examined shortly.

The GRO and the 1841 census

Just as at its outset the provision of medical statistics was not a central feature of the GRO's activities, so it was not obvious that the GRO would become the vehicle for organising the taking of the decennial census, and publishing its results. The first census of England and Wales had been taken in 1801, and it, and all subsequent pre-1841 enumerations, had been organised by John Rickman, a clerk of the House of Commons. Rickman was interested in establishing not only the size of the population but also the demographic factors underlying population change. His enumerations took the form of an enquiry addressed to the local overseers of the poor on the raw numbers of men, women, families and houses in parishes, along with some additional

---

[78] This included occupation; age at admission to friendly society; date of admission; time receiving sickness benefit; date placed on superannuation fund; date of death; date of exclusion; date of leaving society; cause of death; and place of residence at time of death: *Copy of a report and tables, under the directions of the Lords of the Treasury, by the actuary of the National Debt Office*, p. ii.
[79] Ibid.; *Copies of a further report and tables, under the directions of the Lords of the Treasury, by the actuary of the National Debt Office*.
[80] *1st Annual Report of the Registrar of Friendly Societies*, pp. 48–56.

material on ages and occupations. The official returns were only headcounts, although some overseers gathered information on a nominal basis in order to fill out the forms sent to London. In addition, Rickman also sent a set of forms to the local Anglican clergy, on which they were to supply the number of baptisms, marriages and burials in their parishes on an annual basis over the previous decade. Besides providing general information on the nature of, and mechanisms underlying, population trends, these enumerations showed the numbers available for military service, and supported actuarial work.[81]

With the establishment of the GRO the latter enquiry was plainly redundant but Rickman still appears to have been responsible for the initial planning of the 1841 census. In 1836 he had forwarded circulars to the local Church of England clergy so that they could gather information from the parish registers as far back as 1570. He had processed the returns and planned to publish them in order to show long-term population trends. He also drew up a draft Census Bill along the lines of those of his previous censuses, and had looked into the possibility of using the boundaries and officers of the new Poor Law for the enumeration. Rickman seems to have been in charge of the census until mid-June 1840, when he fell ill, and he eventually died in August of that year. It was only in the last week of June that the Home Office discussed with the GRO the possibility of it organising the census.[82] The fact that Lister was the brother-in-law of the previous Home Secretary, Lord John Russell, would probably not have hindered his chances of adding the census to his empire. Lister drafted a new Bill, which was the basis of the first Census Act for the 1841 enumeration.[83] This Act authorised the establishment of a census-taking apparatus that had many similarities to Rickman's censuses. However, the gathering of information on the ground was now to be done by temporary enumerators appointed by the local

---

[81] For the origins and history of the pre-1841 censuses see, Higgs, *Making sense of the census*, pp. 4–7.

[82] PRO: T 1/4573, 10 Feb. 1841, Wm Rickman *et al* to C G Trevelyan [presumably E C Trevelyan, assistant secretary at the Treasury]; Census Returns: Specimens of Forms and Documents (PRO: RG 27): PRO: RG 27/1, p 1.

[83] 3 & 4 Victoria, c. 99.

registrars. They were also to gather a much wider range of information on the individuals of their districts, and this was to be done on one night of the year rather than as previously over a period of weeks. The census was to be a 'snap-shot' of society at one point in time in order to avoid double counting as people moved from parish to parish.

At this initial stage, however, Lister thought that the enumerators would gather the required information by going from house to house and asking questions. He did not favour the use of household schedules, which had been suggested by the Statistical Society of London, because he believed that most householders would be too illiterate to fill them in correctly. Lister only began to countenance the introduction of household schedules after an initial experiment in London had shown how many enumerators would have to be used to collect the information as originally envisaged. The cost of such an exercise would plainly have been prohibitive, and the Treasury obstructive. A supplementary Census Act[84] to authorise the issuing of schedules to householders for them to fill in was rushed through Parliament only two months before the enumeration was due to take place. As in previous censuses, the clergy were required to give the usual information from the parish registers. In the end this data was not collected, although the results of Rickman's analysis of parish register data from 1570 onwards eventually formed part of the *1841 census reports*.[85] These events speak of an ad hoc response to events, rather than a well worked out administrative strategy.

In addition, there is little evidence that William Farr was involved in any of these arrangements. Although he was active on the committee of the Statistical Society of London, which offered advice on the taking of the census, he was not one of the three commissioners responsible for the 1841 enumeration – these being Lister, Edmund Phipps (about whom little is known), and Thomas Vardon, the librarian of the House of Commons. All the surviving correspondence on the organisation

---

[84] 4 & 5 Victoria, c. 7.
[85] PRO: RG 27/1, pp. 6–11; GRO, *Parish registers*.

of the 1841 census appears to have emanated from the Registrar General himself, rather than from his subordinate. Lister died in 1842, before the end of the census abstracting process, and Vardon and Phipps signed the final report.[86]

The role of Thomas Henry Lister

As a consequence of this account, it is necessary to reassess the relative contributions of Lister and Farr to the genesis of the statistical functions of the GRO. Lister has usually been seen as a nonentity, and even such a judicious scholar as John Eyler has been content to follow Finer's assessment of him as a 'flatulent young novelist' and 'decorative headpiece'.[87] During the early years of the GRO, Lister was indeed writing his *Life and administration of Edward, first Earl of Clarendon,* and he had also been appointed Registrar General via the time-honoured tradition of political jobbery. Nor did he have any statistical expertise, as William Farr himself pointed out in one of his early journalistic forays.[88] It is easy to see, therefore, why Lister has been held in such low opinion by historians.

However, as has already been stressed, the GRO at its inception was not primarily concerned with the production of medical and demographic statistics. It is somewhat unfair, therefore, to criticise Lister for not possessing qualities that were not required for the job he was initially asked to undertake. As will be noted shortly, if a criticism can be made of him it was that he spent too much time on establishing the statistical side of the Office's work, and not enough on ensuring the basic administration of its other functions. In truth, without Chadwick's interventions and Lister's interest in statistical work there would probably have been little scope for the employment of a man of the talents of William Farr in the registration service. Historians have perhaps been too ready to assume, with hindsight, that the

---

[86] PRO: RG 27/1, pp. 11–18; PRO: HO 45/146, 10 Jan. 1845, Phipps and Vardon to Phillipps; *1841 census report: abstract of the answers and returns,* p. 72.
[87] Finer, *The life and times of Sir Edwin Chadwick* , p. 143; Eyler, *Victorian social medicine,* p. 46.
[88] Eyler, *Victorian social medicine,* p. 47.

development of the GRO as a centre of scientific medical and demographic research was inevitable, and that intellectual élan counted for everything in the Office's history. Academics can be forgiven, perhaps, for believing in the centrality of intellectual pursuits to history. However, there was nothing inevitable in the development of the GRO as a producer of vital statistics, and the crucial role of the Office's administrative leadership in this process needs to be foregrounded. This is a theme that will be taken up later when considering the development of the GRO in the mid-Victorian period, and the contribution of Major George Graham to it.

3

The expansion of the GRO's statistical functions in the High Victorian period

Property, citizenship and the workings of the registration system

From a perusal of many of the texts that have considered the work of the GRO in the nineteenth and early twentieth centuries, one might well conclude that all its staff were wholly employed on the generation of statistics under the supervision of William Farr, and his successors as the GRO's Superintendents of Statistics. Farr is often seen as the real head of the Office, and sometimes even mistakenly elevated to the post of Registrar General.[1] This is perhaps another example of the tendency of intellectuals to believe that thought always takes precedence over action. In fact, the GRO, and the staff of the local registration service, spent much more of their time on activities underpinning property rights and obligations than on medical or demographic research. So before turning to the expansion of the GRO's statistical functions it might be useful to consider the former in more detail.

In the first full year of its existence, the twelve months ending on 30 June 1838, the clerks in the GRO responsible for compiling the civil registers and indexes processed 399,712 birth certificates, 111,481 marriage certificates, and 335,956 death certificates. By 1888, fifty years later, these figures had risen to

---

[1] Flinn, 'Introduction', p. 27; Finer, *The life and times of Sir Edwin Chadwick*, p. 143; Joyce, *The rule of freedom,* p. 27. This line of argument is not confined to Farr: Margo Conk, for example, elevates T. H. C. Stevenson, the GRO's superintendent of statistics from 1909 to 1931, to the post of Registrar General : Conk, 'Labor statistics in the American and English census', p. 93.

879,868, 203,821, and 510,951 respectively.[2] This volume of work was reflected in the dominance of the Records Department within the staffing of the GRO. As can be seen from Table 3:1, the numbers employed in the section dealing with the compilation of the central register of births, marriages and deaths, and the processing of certificates for the public, always employed the majority of the Office's clerks in the Victorian and Edwardian periods. Over time the proportion of the GRO's clerical staff active in the Statistical Department rose from 8 per cent in 1840 to 27 per cent in 1921, but the Records Department was usually more than double its size. As will be discussed below, the great expansion in the staffing of the Office in the early twentieth century was as much to do with the

Table 3:1. GRO Staffing 1840–1921

Date	Department Correspondence[1]	Accounts	Records	Statistics	Total[2]
1840	10	6	29	4	50
1842 (April)	10	5	50	8	74
1842 (July)	—	—	—	—	87
1843	—	—	—	—	60
1855	—	—	—	—	70
1864	—	—	—	—	78
1866	4	4	50	16	78
1874	—	—	—	—	78
1895	—	11	45	19	78
1905	—	—	—	27	84
1912/1913	—	—	—	—	103
1921	—	9	114	48	172[3]

[1] The Correspondence Branch was amalgamated with Accounts after 1866.
[2] Includes Registrar General, heads of departments and inspectors of registration but excludes industrial grades.
[3] Excludes 38 typists.

Sources: 1840, 1849 Imperial Calendar; 1842 (Apr), RG 29/1, pp. 119–141; 1842 (July), RG 29/1, pp. 202; 1843, RG 29/1, pp. 202; 1855, RG 29/5, p. 385 (pp. 3–4); 1864, RG 29/6, pp. 9–10; 1866, RG 29/2, pp. 54–5; 1874, RG 29/2, p. 184; 1895, T1/8954A/13154; 1905, RG 20/74; 1912–13, T 165/40; 1921, RG 20/81.

[2] GRO, *1st ARRG for the year ending 30 June 1838*, p.19; GRO, *51st ARRG for 1888*, pp. v–viii.

Treasury's provision of extra resources for pensions work, as with new forms of statistical production.

As noted above, the origins of the registration system in England and Wales in the early nineteenth century lay to a considerable extent in the need to secure the citizen's enjoyment of the benefits of ownership of property via the recording of lines of descent. This function was reflected in the day to day activity within the GRO. In his *Annual report* for 1859, George Graham noted that in that year 4,133 successful searches had been made in the registers at Somerset House, the 'greater number' being undertaken by solicitors. There were, of course, a considerable number of such searches that were unsuccessful.[3] As can be seen from Table 3:2, the numbers of searches in the central GRO grew by leaps and bounds. These figures relate only to the numbers consulting the registers in the central reading rooms in London, the use made by the public of the copies of certificates kept by local registrars must also have been extensive. In 1877 Graham noted that, 'the provisions of the Registration Act are becoming more generally known by the legal profession, and increasing numbers of certificates are required for proof of death, as well as for pedigree purposes.'[4]

Table 3:2. Searches in the registers at the central GRO, 1845–1895

Year	Average Annual Searches
1845–9	952
1850–4	1,669
1855–9	3,485
1860–4	7,025
1865–9	11,309
1870–4	18,292
1875	25,407
1885	36,450
1895	53,289

Sources: *38th Annual Report of the Registrar General for 1875* (London, 1877), p. liii; *59th Annual Report of the Registrar General for 1896* (London, 1897), p. xxxvi.

[3] GRO, *22nd ARRG for 1859*, pp. xliv–xlv.
[4] GRO, *38th ARRG for 1875*, p. liii.

The work of the Records Department also took temporal precedence over that of the Statistical Department. Thus, the gap between the collection of mortality data and its publication in the Registrar General's *Annual reports*, sometimes several years in the mid-Victorian period, is explained, in part, by the priority given to the creation of the GRO's property register. Thus, in the *Annual report* for 1860, Graham explained:

> this office was established by Act of Parliament, primarily for the purpose of collecting arranging, paging, examining, correcting, binding, and indexing the certified copies of the English registers, and of supplying stamped certificates of births, deaths, and marriages to all persons who may apply for them; that the certified copies are received quarterly, but the returns are not completed until nearly three months after the end of each quarter; and that the preparatory duties which I have mentioned, and which occupy a majority of the gentlemen engaged in the office during a period of eight months after the quarterly arrivals have begun, must be performed in respect to each volume before it can pass into the hands of the statistical clerks.[5]

The burdens placed upon the central GRO for research facilities and official copies of certificates by the requirements of the legal profession and the public were large, and constantly growing. The expanding numbers of citizens wishing to consult the registers and indexes was such that accommodation for search rooms, and the number of porters employed, in Somerset House had to be constantly increased. As early as 1846 Graham was complaining to the Treasury about the lack of accommodation for the bound volumes of certificates and the public. Some of the registers he warned might have to be stored in underground rooms, thus increasing the work of the messengers who had to fetch them.[6] In 1865 Graham requested the authority to employ more boy porters since he found it, 'Desirable that additional assistance should be afforded, 50 or 60 persons attending

---

[5] GRO, *23rd ARRG for 1860*, p. xliv.
[6] PRO: RG 29/1, p. 261.

here sometimes in the course of one day.'[7] The following year he again complained about the restricted facilities for the public and suggested the removal of the Society of Antiquaries and other learned societies from Somerset House to make more room.[8] This had come to pass by 1875, and the 'industrial' complement of the Office was now eight messengers, four porters, and two assistant boy porters under the office keeper, who 'assist the public, and are instrumental in obtaining successful searches and in tracing pedigrees and genealogies'. By 1881 the search rooms, and the hall leading to them, employed 10 designated search room attendants, and fees collected from the public reached nearly £5,000 per annum.[9] Thirteen years later, in 1894, there were 11 search room attendants paid on a scale from £75 to £115.[10] As can be seen from Table 3:1, at this date the entire clerical complement of the Statistical Department was only 19, amongst which the assistant clerks acting as statistical abstractors started on a pay scale of only £55 per annum.[11]

Much the same argument with respect to the priority of administrative over scientific functions could be made in the case of the local registrars. Although these officers frequently supplied local medical officers of health with mortality statistics in order to facilitate medical and sanitary work within their districts[12], this was hardly the core of their activities. At the inception of the registration service there had been an initial surge of interest amongst members of the medical profession in acting as local registrars. This may have been due to Thomas Lister's exhortation to doctors via the London societies in 1838 to support the new registration system. Certainly, Lister was able to claim in the first *Annual report* in 1839 that out of 2,193 registrars appointed at that date, 416 were medical officers of local Poor Law unions, and 111 were other members of the medical

[7] PRO: RG 29/2, p. 36.
[8] PRO: RG 29/2, p. 54.
[9] PRO: RG 29/2, p. 310.
[10] PRO: RG 29/3, p. 159a.
[11] PRO: RG 29/3, p. 215.
[12] Lewes, 'The GRO and the provinces'.

profession.[13] By 1863, however, George Graham was downplaying the medical input into the registration service, noting that:

> Undoubtedly there are medical men who act as registrars in England, and who perform the duty in the most satisfactory manner; the names of many might be mentioned; but it must be acknowledged that, nevertheless, that the habits of a surgeon in good practice *tend* to disqualify him for the sedentary mechanical office of entering after careful enquiry, with measured movement, and in a fine round hand, a number of particulars in the forms of a manuscript book. The 'prescription hand' is notoriously ill adapted to this purpose.[14]

In the early years of the twentieth century, members of the medical profession had almost ceased to enter the registration service. Thus, of the 673 registrars of births and deaths appointed in England and Wales in the period 1907 to 1914, only nine were doctors, whilst 265 were Poor Law relieving officers, and 115 other local authority employees.[15] The local registrars were so ignorant of medical terms by this date that the GRO had to issue them with a special index of causes of death.[16] From 1930, under the operation of the 1929 Local Government Act, the registration service was gradually transformed from a fee-paid to a salaried service. By 1950, out of 1,752 registrars of births and death only 35 were not salaried, and the medical component amongst the part-time officers had all but disappeared.[17] Registration at the local level was an administrative rather than a scientific activity.

---

[13] GRO, *1st ARRG for Year ending 30 June 1838*, p. 4.
[14] GRO, *24th ARRG for 1861*, p. xliv.
[15] PRO: RG 21/108, Registration officers vacancies and appointments registers, 1906–1939, pp. 31–55.
[16] PRO: RG 29/4, Registrar General's letter to the Treasury of 20 November 1911.
[17] *Registrar General's Statistical Review for the Five Years 1946–1950: Text, Civil*, pp. 158–60.

The rise and fall and rise and fall of the *Annual report of the Registrar General*

However, despite the contemporary importance of the GRO's work in recording individual vital events for civil and legal purposes, it is the aggregate statistical work that has preoccupied historical accounts. In this the present work is no exception. Nevertheless, not withstanding the extensive use made of the GRO's reports, and the general high esteem in which they are held, there have been comparatively few attempts to examine their changing form and structure, or the history of their production.[18] In the absence of a collected series of these publications, and adequate indexes to them, this scholarly omission is perhaps understandable. However, the lack of an overall appreciation of the nature and rhythms of the GRO's publishing endeavours disguises some of the important features of the Office's scientific development, and can lead to confusion. In order to follow and explain the development of the GRO's statistical work, it is necessary to make a preliminary attempt to sketch out the gross features of this textual history from 1839 to the inter-war period. A more detailed listing of the Office's main non-census publications can be found in Appendix 1.

The centre-piece of the GRO's publishing endeavours for most of the period under consideration here was the *Annual report of the Registrar General*, which discussed the registration, demographic and medical data abstracted for a single twelve month period. This report might appear up to 40 months after the year upon which it commented, although a twelve-month gap was more usual. As can be seen from Figure 3:1, these delays tended to be most extensive in the early years of the GRO's existence but nearly always grew in the first half of each decade, before declining in the second half. As already noted, a delay in publication was inevitable given the priority given to the creation of the registers of births, marriages and deaths. However, variation in the length of the delay was probably

---

[18] Notable exceptions have been: Szreter, 'The GRO and the public health movement', pp. 438–9, 454–5; Hardy, 'Death is the end of all disease', pp. 473–4. An earlier version of the arguments in this section can be found in Higgs, 'The Annual Report of the Registrar General'.

explained in terms of the large amounts of extra work that the organisation, taking and reporting of the census created for the Office every ten years. In 1915, for example, the then Registrar General explained the delay in publishing the *Annual reports* for 1911 and 1912 partly in terms of the 'extra pressure on the Department owing to the census – a cause which has regularly produced the same result in the past . . .'.[19] The 1841 census was, of course, the first the GRO organised, and that of 1851 was in many ways its most ambitious, and this may explain, in part, the extreme delays in producing the *Annual reports* in these decades.

The GRO attempted to get round this problem in the 1840s and 1850s by producing a shorter abstract that was laid before Parliament, and by publishing the full report with an explanatory text separately at a later date. Thus, an abstract relating to the births, marriages and deaths in 1848 had been laid before Parliament in 1850[20], but the full report was not completed until December 1851, and not published by the Stationery Office until 1852.[21] This means, at least for some years in the nineteenth century, that the *Annual reports* found amongst the published *Parliamentary Papers* are not the full texts, a phenomenon which can cause confusion.[22] On occasions, as in 1886 and 1894, the GRO attempted to catch up by producing two reports in one year.[23]

From its second appearance, in 1840, the text of the *Annual report* was usually in two parts – the Registrar General's own report proper, and the 'Letter to the Registrar General' under the signature of the Office's Superintendent of Statistics, each with relevant tables. The former was usually a broad-brush reflection on the trends in vital events, with digressions on such matters as the ability of people to sign the marriage register; the relationship between marriage rates and the economy; illegiti-

---

[19] GRO, *76th ARRG for 1913*, p. viii.
[20] *Eleventh Annual Report of the Registrar General for 1848*.
[21] GRO, *11th ARRG for 1848*, p. i.
[22] Hardy alludes to this problem: Hardy, 'Death is the end of all disease', p. 474, n. 7. In the present work all references are made to the complete, independently published versions of the *Annual reports* rather than to Parliamentary Papers, although after that for 1851 they are identical.
[23] GRO, *47th ARRG for 1884*; GRO, *48th ARRG for 1885*; GRO, *55th ARRG for 1892*; GRO, *56th ARRG for 1893*.

Figure 3:1 Delays in publishing the Annual Reports of the Registrar General

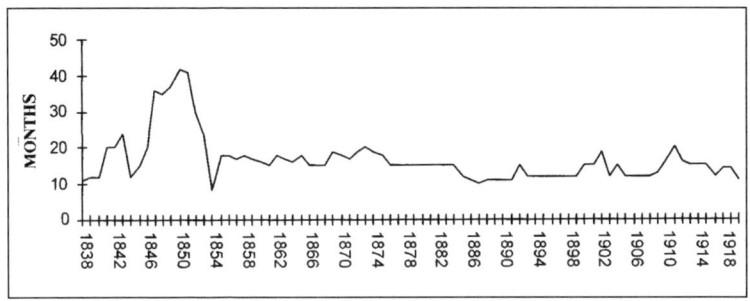

Source: Annual report of the Registrar General

macy levels; the effect of the weather on death rates; and so on. For much of the 1850s and 1860s the Registrar General's personal report also included a summary of the quarterly reports on mortality in the various registration districts, which the GRO began compiling in the 1840s. However, this summary disappeared in the *Annual reports* after that for 1873.[24]

The Superintendent of Statistic's 'Letter', which is frequently confused with the *Annual report* as a whole, was usually concerned with mortality, or life tables, although Farr occasionally digressed into reports on the international statistical congresses he had attended, and the like. Compared to the Registrar General's report, this was a much less constant feature of the *Annual report*, as can be seen from Figure 3:2.[25] In many years the Registrar General's own report was larger, and even during Farr's period of office the Letter might not appear at all.[26] As Simon Szreter notes, the Letter signed by the Statistical

---

[24] GRO, *37th ARRG for 1874*.

[25] This graph relates solely to the textual commentary on the published tables but this can be taken as indicative of the GRO's commitment to a public propaganda role.

[26] This appears to have been especially the case in the 1840s when the 7th, 9th, 10th *Annual reports* do not contain a Letter from the superintendent of statistics. These problems are noted by Hardy, 'Death is the end of all disease', p. 474. She claims, however, that there no published cause of death information published between 1848 and 1855, which is incorrect: see GRO, *12th ARRG for 1849*, pp. 252–69; GRO, *13th ARRG for 1850*, pp. 150–67; GRO, *14th ARRG for 1851*, pp. 120–37; GRO, *15th ARRG for 1852*, pp. 122–41; GRO, *16th ARRG for 1853*, pp. 124–39: GRO, *17th ARRG for 1854*, pp. 120–139. There are also summaries of the quarterly returns in all these reports.

Superintendent disappeared in the *Report* for 1879, only returning in that for 1901.[27] However, during the intervening period its content was incorporated into the Registrar General's own report under the heading 'Registered Causes of Death'. The *Reports* of the last two decades of the Victorian period were, on the whole, slighter documents than those that preceded and followed them. With the *Annual report* for 1901 the division between the Registrar General's report and the Letter reasserted itself.[28] From the *Annual report* for 1909 onwards, however, the Registrar General's own signed report became a mere introduction of a few pages to the Superintendent's Letter, now renamed the 'Review of Vital Statistics'.[29] The *Annual report* for 1920 was in fact an unsigned version of the Superintendent of Statistic's Review.[30] The following year this anonymous format continued but the whole report was now re-titled *The Registrar General's statistical review for 1921* – the first in a new series.[31] The *Statistical review,* unlike its predecessor, was no longer placed before Parliament as a parliamentary paper.

The *Annual reports* did not, of course, exhaust the GRO's published output.[32] By 1840 the GRO was collecting cause of death data from the Metropolitan registrars on a weekly basis, and publishing an abstract of the information. From the mid-1860s similar material was published for ten large provincial towns, and the numbers of places so covered increased during the course of Victoria's reign. As already noted, from 1842 onwards the Office published quarterly reports on general mortality rates in 114 of the most populous districts of the country. A new quarterly series covering all registration districts began in 1849, and in 1870 these were upgraded to give cause of death. In 1864, Farr published a supplement to the *Annual report* in which he summarised the mortality experience of

---

[27] Szreter, 'The GRO and the public health movement', p. 454; GRO, *42nd ARRG for 1879.*
[28] GRO, *64th ARRG for 1901.*
[29] GRO, *72nd ARRG for 1909.*
[30] GRO, *83rd ARRG for 1920.*
[31] GRO, *Registrar General's Statistical Review for 1921.*
[32] Szreter, 'The GRO and the public health movement', pp. 438–9.

Figure 3:2 Pages of text in the Annual Report of Registrar General

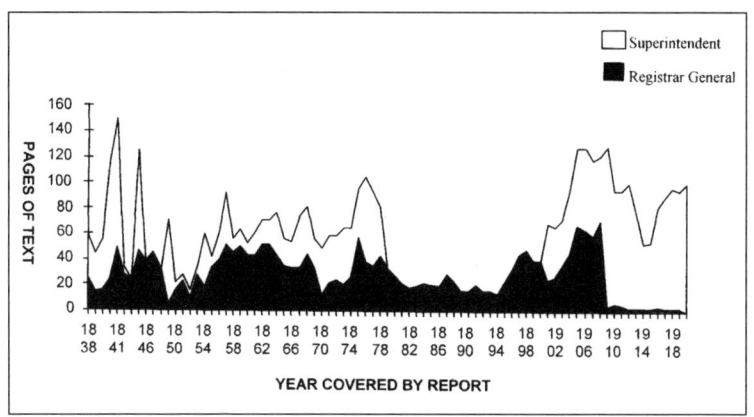

Registrar General = Registrar General's own report
Superintendent = Letter of the Supreintendent of Statistics

Source: Annual reports of the Registrar General

England and Wales in the two decades 1841–50 and 1851–60.[33] The *Decennial supplement* then became a standard periodic feature of the GRO's output in the period covered by this work (see Appendix 1).

The history of the *Annual reports* can thus be divided into four, fairly distinct, periods:

(i) the joint publication of reports by Lister and Farr, and by Graham and Farr, covering the period 1839 to 1879;
(ii) the last twenty years of the nineteenth century, which saw the absorption of the Letter into the Registrar General's own report, coupled with a general decline in the GRO's output;
(iii) the rebirth of the separate Superintendent of Statistics' Letter at the beginning of the twentieth century; an expansion of publishing activity in the *Reports*; and the absorption of the Registrar General's contribution into that of his statistical subordinate.

[33] GRO, *Supplement to the 25th annual report of the Registrar General*.

This development terminated (iv) with the imposition of anonymity and a new title in the 1920s. The pattern of expansion, retrenchment, revival and sudden termination in the history of the *Annual report* reflects distinct phases in the administrative and political history of the GRO. The rest of this chapter will analyse the first phase of this development, from 1839 to 1879, whilst later chapters will deal with the subsequent periods.

Farr, medical statistics and the nature of the State

As already noted, the development of the GRO during its first phase of expansion between 1839 and 1879 has tended to be seen in terms of the interests and career of William Farr. This reflects the manner in which the GRO's history has been based almost exclusively on its publications, and these approached, in turn, in terms of the history of science, rather than from an institutional standpoint. Thus, John Eyler's magisterial treatment of the GRO's first Superintendent of Statistics, subtitled, 'The ideas and methods of William Farr', is essentially an intellectual biography.[34] The GRO is seen here almost as the institutional support for Farr's personal research projects. To a certain extent, Simon Szreter continues this line of argument in his consideration of the work of the later Superintendents of Statistics on the effects of occupation and social class on morbidity and fertility.[35] The Office's work is seen as part of the ongoing intellectual debate within the medical and demographic sciences. Indeed, the portrait that emerges of the GRO in the late Victorian and Edwardian periods appears to bear more than a passing resemblance to a modern university research unit.

William Farr plainly had both a scientific and a political project to forward in his work within the Office. As Eyler has shown, Farr used the cause of death data gathered in the process of civil registration to elucidate the temporal and spatial patterns of epidemic and other diseases, and to put forward important hypotheses regarding their causes. His development of disease

---

[34] Eyler, *Victorian social medicine, passim.*
[35] Szreter, *Fertility, class and gender in Britain,* pp. 76–282.

classification systems (nosologies), facilitated the statistical analysis of mortality, whilst his work on the relationship between mortality, elevation of place of residence, water supplies and density of population, added considerably to the scientific understanding of the aetiology of disease. Farr was also a leading light of the great institutions of Victorian social science, including the British Association for the Advancement of Science, and the London, later Royal, Statistical Society, of which he was president in 1871 and 1872.[36] He was also a member, if an independently-minded one, of the circle of reformers around Florence Nightingale.[37]

Farr believed that the key to understanding differential death-rates between the countryside and the town, and between various areas within towns, was population density. The higher this was, the greater the problems associated with removing human effluent, which was in turn the prime source of the 'zymots' which led to disease. 'Zymots' was Farr's term for hypothetical chemical pathogens, which were supposed to have poisoned the blood. Thus, the cause of cholera was taken to be an unknown chemical, dubbed 'cholerine', in the surroundings of the deceased. Indeed, the whole structure of the mid-Victorian censuses can be seen in terms of a survey of the factors contributing to the growth of population density, and thus to the health of the people. Hence questions on the numbers of houses and people in defined areas; on the determinants of family formation and reproduction, such as the age, sex and marital status of the population; and patterns of migration. These essentially environmentalist concepts of disease were linked to some extent with contemporary belief in the danger of 'miasma', but more closely with the chemical theories of Justus von Liebig.[38]

But, as Szreter has rightly stressed, Farr also used the differences in the published rates of deaths per thousand in the various registration districts to highlight local sanitary failings. The 1848 Public Health Act contained a provision that local authorities

---

[36] Eyler, *Victorian social medicine*, pp. 66–122.
[37] Ibid., pp. 155–189.
[38] Ibid., pp. 100–5; Higgs, 'Diseases, febrile poisons, and statistics'.

were to be compelled to establish health boards to implement local sanitary reform if their annual mortality rates were found by the GRO to be above 23 per thousand. In the course of the 1850s Farr developed the concept of the 'Healthy Districts' mortality experience, based originally on a set of 63 rural registration districts whose crude death rate was below 17 per thousand. This was then advocated as the desirable national standard, and deaths in excess of this were designated as preventable. The GRO shamed local authorities into sanitary reforms by publishing calculations indicating that tens of thousands of deaths in their cities would never have occurred if only the sanitary conditions of these districts approximated to those of the Healthy Districts.[39] It should be noted, however, that this strategy was not entirely without precedent. As early as 1844, Edwin Chadwick was using data obtained from the GRO to show the differences in the numbers of births and deaths in London which would have resulted if the Metropolis had the same mortality regime as Herefordshire.[40]

In his writing Farr used impassioned rhetoric to make an impact, as can be gauged from the following extract from the quarterly returns in the *Annual report* for 1853:

> What no sceptical philosopher would have dared to propose as an experiment, what no haughty conqueror ever condemned the inhabitants of a subjugated city to endure – this fine English town on the Tyne – the centre of the coal trade – of intelligence of every kind – and of engineering knowledge – has done and suffered. All the excreta, which are thrown into the streets or water closets, are washed down the acclivities of the streets into the river; the fermenting mass is driven up and down by the tides, and has thence since July been pumped by the engine at Elswick all over the town through the water pipes for domestic uses: it has been used for ablution, it has been washed over the floors, it has been drunk as a beverage by many of the children and the wives,

---

[39] Szreter, 'The GRO and the public health movement', pp. 439–40.
[40] Chadwick, 'On the best modes of representing accurately, by statistical returns, the duration of life, pp. 16–17.

The expansion of the GRO's statistical functions

as well as large numbers of the higher and middle as well as the working men of the town. This sad fact in the history of Newcastle will be remembered when the loss of 1500 lives [by cholera], by which it was followed, is forgotten.[41]

These stylistic flourishes, and the Office's general deployment of crude mortality statistics, did not go unchallenged, being the targets of attacks on the GRO's methods by local doctors and medical officers of health (MOHs), such as Henry Rumsey and Henry Letheby.[42]

The GRO's mass dissemination of thousands of free copies of its own reports to registration officers, coroners, learned societies, medical practitioners, mechanical institutes, and private individuals, led to the formation of a well-informed and vociferous public opinion.[43] The use of public funds to spread public-health propaganda in this manner was a feature of this period, as can be seen from the free distribution of thousands of copies of Edwin Chadwick's *Report on the sanitary condition of the labouring population of Great Britain* in the 1840s.[44] These propaganda activities, and the statistical methods used by Farr became a model for the activities of local statistical societies and MOHs. As already noted above, none of these statistical developments flowed necessarily from the provisions of the 1836 Registration Act, rather they were the result of administrative action. The early decades of Victoria's reign were a period when energetic civil servants, especially experts with specialist knowledge, could still mould the form and activities of their departments independently of Parliament.[45] Officials, such as Chadwick, John Simon at the Medical Department of the Privy Council, and Roland Hill at the Post Office, were frequently prepared to appeal directly to the public over the heads of their departmental superiors, and even of their ministers, in pursuit of their own objectives.[46]

[41] GRO, *16th ARRG for 1853*, p. 38.
[42] Eyler, 'Mortality, statistics and Victorian health policy'; Mooney, 'Professionalization in public health', p. 63.
[43] PRO: RG 29/1, p. 551.
[44] Flinn, 'Introduction', pp. 55–7.
[45] MacLeod, 'Introduction', pp. 9–10; Eyler, *Victorian social medicine*, p. 51.
[46] Clark, 'Statesmen in disguise'.

Farr was also fortunate in working in a department over which there was little active ministerial supervision, which gave him a freer hand to drive the GRO's statistical work in the direction of his own choosing. As already noted, the revisions to the original 1836 Registration Bill had left the GRO in a position of relative constitutional independence. In theory the Office came under the Home Office until 1871 but the Registrar General supervised the workings of the registration system, and dealt directly with the Treasury over recruitment and staffing. The lack of central control can be seen in the nature of the material contained in the Home Office's entry books of correspondence with the Office. Throughout the period 1836 to 1871 these indicate little discussion of policy, being confined to low-level procedural matters.[47] The lack of effective structures for ministerial and Parliamentary control was not an uncommon feature of many of the new Civil Service bodies being created in this period.[48]

Farr and the nature of statistics

It should be noted that the work of William Farr was not concerned with statistics in the modern probabalistic sense of that term. As John Eyler has argued, for men such as William Farr and his early Victorian contemporaries 'statistics' was not a branch of mathematics but 'the science of States – the science of men living in political communities . . .'.[49] This meant the comparison of nations rather than the study of normal distributions. Farr rarely used calculus in his published work and, with the exception of his life tables, he seldom treated phenomena as continuous functions. He never spoke explicitly of data distrib-

---

[47] PRO: HO 34/1–25. The subjects covered included: prosecutions for non-registration; redirecting correspondence; the Home Office forwarding statistical information from abroad; the GRO keeping the Home Office informed of developments, especially with respect to negotiations with the Treasury; the formal Home Office sanction of the employment of people in the Census Office; the Home Office asking for the Registrar General's comments on relevant bills; and so on

[48] Clark, 'Statesmen in disguise', pp. 29–30; Eyler, *Victorian social medicine,* p. 51.

[49] This is an excerpt from William Farr's presidential address to the Statistical Society of London, 1872, quoted in Eyler, *Victorian social medicine,* p. 28.

ution, or used concepts of statistical error. Farr's stock in trade were tables of raw aggregates and simple ratios showing local variations of the rate of deaths per thousand, rather than the use of equations or probabilities. Fred Lewes has argued, however, that Farr's ideas on the relationship between the incidence of cholera and elevation show him moving towards the concepts of continuous functions and general linear equations.[50] Theodore Porter is particularly scathing about the manner in which early to mid nineteenth-century statisticians were content to compare rates of natality, mortality, marriage, crime and suicide, describing this as 'a strategy especially suitable for deeply ambitious but fundamentally sensible persons of moderate ability'.[51] This is perhaps a little harsh since, as will be discussed shortly, the GRO needed to produce simple statistics that could be calculated by poorly trained clerks, and be readily understood by a generally innumerate British public.

The scope of Farr's overall statistical project can be seen in his report on the proceedings of the first International Statistical Congress in Brussels in 1853. In this he sketched an outline for a general framework for national statistics to facilitate international comparisons. Each state was to establish machinery to collect information on the following subjects:

1. The revenue and expenditure well classified
2. The civil and military establishments. The office and expense of local and municipal government.
3. The census of population at the same, and at equal, or at least corresponding intervals of time.
4. The registration of marriages, births and deaths and causes of deaths.
5. The emigration and immigration of population.
6. The churches, scientific societies, schools, public institutions of various kinds, connected with the moral and intellectual progress of the people.

---

[50] Eyler, *Victorian social medicine*, pp. 66–8; Lewes, 'William Farr and the communication of cholera', p. 10.
[51] Porter, The *rise of statistical thinking*, p. 46.

7. Friendly societies, savings banks, insurance societies, institutions for the relief of the poor, charities and social institutions of various kinds
8. Crimes and punishments.
9. Surveys, maps, descriptions and measurements of the territory; indications of the superficial and subterraneous contents of the land; observations on the temperatures, rain and other elements of climate
10. Statistics of agriculture}
11. Statistics of industry}   Variously sub-divided
12. Statistics of commerce}[52]

Similarly, in June 1855 Graham forwarded a letter from Farr to the Treasury regarding his request to attend the second International Statistical Congress in Paris. The stated aim of this Congress was to establish plans, 'for classifying and arranging the principal facts respecting, population, trades, manufactures, commerce, education, crime, etc., etc., in such a manner as to facilitate their comparison, and the deduction of general results from the national returns'.[53] Farr's report to the International Statistical Congress at The Hague in 1869 was not medical in content but a lengthy discussion of the progress of metric weights and measures in England, and the development of the coinage.[54] Such a concept of statistics had more to do with the *encyclopedists* of the French Enlightenment than with modern statistical theory. As Abrams has argued, this manner of approaching the collection of information was eminently suited to a period of political consensus, when the role of the social sciences was to gather data to inform administrative activity, rather than to develop new analytical tools.[55]

For Farr quantification was a means of summarising aggregate data, in which human beings:

[52] PRO: T 1/5828B/20825.
[53] PRO: RG 29/1, p. 466.
[54] GRO, *31st ARRG for 1868*, pp. 235–84.
[55] Abrams, *The origins of British sociology 1834–1914*, pp. 13–15. According to MacKenzie, in the first 50 years from its foundation in 1834 only 2% of the papers read to the Statistical Society of London dealt with statistical methods, MacKenzie, *Statistics in Britain*, p. 9

appear divested of all colour, form, character, passion, and the infinite individualities of life: by abstraction they are reduced to mere units undergoing changes as purely physical as the setting stars of astronomy or the decomposing atoms of chemistry; and as in those sciences so in this, the elementary facts observed in their various relations to time and place will shed new light on the more complicated phenomena of national life.[56]

The reduction of human beings to similar social atoms capable of aggregation plainly fitted into a particular liberal concept of the citizen body. Since citizens were the same, it was the effects of the environment ('time and place') working on differing subpopulations that needed to be analysed. From being the expression of disharmony in the unique constitution of the individual, the locus of disease was now only incidental in humans, it became a matter of 'places, not persons' as Southwood Smith put it.[57] Such positivistic assertions, that one can use quantification to remove all contingency, and also experience unmediated 'reality', was belied by the ideological assumptions that underlay much of the GRO's statistical work.[58]

Farr's overarching concept of statistics was undermined in the late Victorian and Edwardian period by the emergence of modern mathematical statistics based on the work of men such as Francis Galton and Karl Pearson.[59] Eugenicists such as Galton were not concerned with reducing differences between populations to averages, but in measuring the differences that existed within populations. All citizens were not equal units. This led to the development of the modern statistical methods for measuring dispersion, distribution and probabilities within data. Inevitably, Farr and Galton clashed over the latter's criticism of the lack of mathematical rigour in the work of the statistical section of the British Association for the Advancement of

[56] GRO, *Supplement to the 35th ARRG*, p. iii.
[57] Hamlin, *Public health and social justice,* p. 119.
[58] Higgs, 'The linguistic construction of social and medical'; Higgs, 'The General Register Office and the tabulation of data'.
[59] For this development see, for example, Hilts, *Statist and statistician*; MacKenzie, *Statistics in Britain.*

Science. Under Farr's direction the Association's Anthropometric Committee, for example, used simple relationships such as the comparison of mean values, whilst under Galton its work included the use of probability theory, normal distributions, and the quartile and decile as well as the mean. By the time of his death in 1883, Farr's grandiose concept of statistics was already becoming outdated.[60] The old comparative science of states was breaking up into a number of distinct disciplines – economics, sociology, mathematical statistics, and eugenics.[61] The GRO continued to employ Farr's traditional methodologies well into the next century, with dire consequences for its intellectual standing in Whitehall, as will be discussed in later chapters.

The skills base of the Victorian GRO

Such considerations raise the question of the normal content of the work of the staff of the GRO's Statistical Department, and the nature of their skills. In general, it is necessary to emphasise the routine and mechanical aspects of their activities rather than their scientific expertise. Although some of the Department's staff appear to have had a medical background[62], most of the men working there had little in the way of scientific or statistical training.

One of the consequences of the Northcote-Trevelyan reforms of the mid-Victorian Civil Service was to divide civil servants into those performing 'intellectual' work, and those whose activities were 'mechanical.[63] The pay and conditions of these two groups of clerks were to be kept quite separate, and the entrance examinations for each division assumed different levels of educational attainment. This distinction was intended to be one made within the established Civil service. In the mid-Victorian period, however, the lack of higher grade civil

---

[60] Eyler, *Victorian social medicine*, p. 28.
[61] Abrams, *The origins of British sociology*, pp. 78–100. For the distinction between the statist and the statistician see also, Desrosières, *La politique des grands nombres*, pp. 20–3.
[62] PRO: RG 29/1, pp. 278–9; RG 29/2, p. 327; RG 29/5, p. 296.
[63] Wright, *Treasury control of the Civil Service*, pp. 98–122.

servants meant that the lower division clerks often had to be used for 'intellectual' work, whilst temporary and unestablished clerks, copyists and writers were hired for routine activities. In 1895, for example, the GRO's Statistical Department contained only two clerks in the higher division out of a clerical compliment of 18.[64]

The Civil Service entrance examinations did not include any specifically medical material, or necessarily imply a knowledge of advanced statistical methods. In the early 1870s, for example, those taking examinations under Regulation I ('intellectual' work) were expected to answer questions on 'pure and mixed' mathematics, zoology and botany but nothing on strictly medical subjects. Those entering the examinations under Regulation II ('mechanical' work) only had to answer questions on arithmetic, and there was no specific scientific component in the paper.[65] As late as 1906, A. L. Bowley was complaining in his presidential address to Section F of the British Association that mathematical statistics was absent from the syllabuses of both Oxford and Cambridge, and from the Civil Service examinations.[66] When George Graham had been asked by the Treasury in 1859 to indicate what he thought should be in the examinations, he replied that clerks destined for the GRO should cover writing from dictation, arithmetic (including vulgar and decimal fractions), English composition, the ability to precis, geography, and English history. On the other hand, his indexers, transcribers, sorters and statistical abstractors only needed to be tested on handwriting, orthography, accuracy in copying, and arithmetic.[67]

Once staff entered the Statistical Department they plainly had to undergo a considerable period of training 'on the job', especially in order to learn the rudiments of medical nosologies. In the 1880s this training was said to require 'many month'.[68] Such induction, and the general supervision of the clerks' work,

[64] PRO: T 1/8954A/13154. The rest of the clerical staff comprised seven second division clerks, six statistical abstractors, a copyist, and two boy clerks.
[65] Ibid., p. 98.
[66] Bowley, 'The importance of scientific method', pp. 204–7.
[67] PRO: T 1/6233A/20316: letter from Graham dated 29 January 1859.
[68] PRO: RG 29/3, p. 64.

appears to have taken up a considerable proportion of the time of the senior officers of the department. Thus, in 1842, Farr's duties, besides drawing up the annual Letter to the Registrar General, were said to include ensuring, 'that the abstracts and tables which the Registrar General directs to be made are correctly performed'; classifying the causes of death; suggesting 'the various tabular arrangements of the facts'; and giving his clerks 'the methods of performing the various calculations . . .'.[69] As Eyler has noted, Farr was more interested in obtaining decent raw data, and securing the creation of reliable basic indices, than in advanced statistical methodologies.[70] It was the gradual perfection of the expertise of his clerks in these duties as time went by which allowed Farr to spend more of his time on unofficial statistical projects. This expertise was essentially in the manipulation of data, rather than in scientific or mathematical abstraction.[71]

Nor did the staff of the GRO themselves think it important that the Office should have medical expertise. When William Farr retired in 1879, almost the entire senior staff of the Office above the grade of lower division clerk, including Noel Humphreys, Farr's biographer, wrote to the new Registrar General, Brydges Henniker, pointing out:

1) That the duties of the superintendent of the Statistical Department of this office are essentially statistical rather than medical.
2) That as the GRO now stands in intimate relationship with the Local Government Board, this Department can have the advice of the medical staff of that Board on any medical question that may arise for the decision of yourself or the superintendent of statistics.[72]

Plainly the clerks were frightened that the requirement for

---

[69] PRO: RG 29/1, p. 133.
[70] Eyler, *Victorian social medicine*, pp. 66–7.
[71] For a description of the sort of work performed by the clerks, see PRO: T 1/8954A/13154; PRO: RG 29/2, p. 354.
[72] PRO: RG 29/2, pp. 283–4.

The expansion of the GRO's statistical functions

medical expertise would lead to an outside appointment, as indeed transpired. However, this hardly indicates the existence of a strong scientific culture within the Department. As will be discussed below, the appropriate balance between statistical and medical expertise within the GRO was to become a crucial issue when the Office passed under the control of the Ministry of Health in 1919.

One should not think, therefore, of the Victorian GRO's Statistical Department in terms of the activities of an academic research unit. Much of its work was routine, repetitive and profoundly dull – data management, and its supervision, rather than scientific analysis.

The role of Major George Graham

Given these limitations, Farr's output plainly represented a great intellectual and technical achievement but this does not explain how the GRO was able to pursued the Treasury to increase its resources to undertake these activities. Why was the Treasury, which had almost blocked Farr's appointment in 1838, prepared to sanction a fourfold increase in the clerical staff of the Statistical Department in the period 1840 to 1866 (see Table 3:1)? That the Office was able to gain the support of the Treasury was not due directly to Farr but to the abilities and energy of his superior, Major George Graham, the Registrar General from 1842 to 1879, who has been overshadowed almost completely by his subordinate. Those historians who have even noted his existence have tended to regard him as either a weak figure[73], or as a capable but dull administrator who contented himself with providing Farr with the tools for his statistical work. On the basis of two letters from Graham which survive in Farr's papers, John Eyler concludes that Graham was, 'slow to make changes, reluctant to put pressure on the government, and interested in guaranteeing the accomplishment of routine tasks.'[74] However, this is hardly the figure who emerges from an analysis of the extant administrative record, and a good case can

[73] Lambert, *Sir John Simon* , p. 419.
[74] Eyler, *Victorian social medicine,* p 49.

be made for the success, and much of the élan, of the GRO in the high Victorian period as being the product of Graham's character and abilities. Indeed, the intellectual history of the GRO needs to be written in terms of the working relationship that existed between the head of the Office and his Statistical Superintendent – when both were dynamic the GRO could be a statistical powerhouse, but when one or other was lacking the institution could easily descend into mediocrity.

George Graham was born in 1801, the younger son of an important landed family in Cumberland, and the brother of Sir James Graham, the Conservative Home Secretary from 1841 to 1846. He entered the East India Company Service, and retired with the rank of Major in 1831, having been military secretary in Bombay between 1828 and 1830. Graham became private secretary to his brother on the latter's appointment as Home Secretary in the Peel Ministry, and was given the post of Registrar General on the death of Thomas Lister in 1842, presumably via his kinsman's influence.[75] Such personal patronage was not unusual in this period, and Sir James Graham was indeed noted for his practice of putting efficiency before political jobbery in his appointments.[76] Graham was thus a member of the political and landed establishment, and used to dealing with his political masters on terms of social equality.

Rather than a weak or dull man, Graham's correspondence reveals him as a determined, fiery character, almost to the point of recklessness. In this he was similar to his elder brother, whose political career was marred, according to the *Dictionary of national biography*, by his being 'so little conciliatory in manner and so rash in utterance.' In 1863, for example, when the GRO's chief clerk of accounts died, the Treasury suggested amalgamating the accounts, correspondence and stores departments, and promoting Farr to chief clerk in order to save money.[77] Graham immediately wrote a letter defending the existing structure of the Office and his proposed promotions consequential upon the chief clerk's death. The tone of this missive, which is worth

---

[75] *Dictionary of National Biography*, 'Sir James Graham'; Nissel, *People count*, p. 147.
[76] Cohen, *The growth of the British Civil Service*, pp. 75–6.
[77] PRO: RG 29/5, pp. 467–9.

quoting at some length, is indicative of Graham's style and forceful personality:

> many persons have knowledge of many things, or think that they have. I profess to know only one thing, the proper management of the GRO; and I think that their Lordships will not do wrong if they were to give me credit for knowledge upon this one single point attainable without any peculiar talent or science and acquired after the drudgery of attending to it since 1842. After 21 years of devotion to this department and with a better and more complete knowledge of each individual in it and of the duties assigned to each individual than their Lordships in Downing Street can possibly possess, endeavouring to administer strict justice and not to sheer partiality, favour or affection, striving only for the public good, if I am not allowed to promote those with whom I have been in daily intercourse and whom I have personally ascertained to be best fitted and most deserving, I must have forfeited their Lordships' confidence ....[78]

The Treasury backed off and agreed almost immediately to Graham's recommended promotions.[79]

Nor were the recipients of such missives confined to members of the Civil Service. Graham was one of the few heads of departments to oppose the introduction of the distinction between 'intellectual' and 'mechanical' work, which were enshrined in the Northcote-Trevelyan reforms of the Civil Service.[80] Graham objected to the reduction in the terms and conditions of many of his staff implied in this division, and in 1871 wrote a blistering letter to Robert Lowe, the Chancellor of the Exchequer. This concluded sarcastically:

> After 30 years in this office I find my duties merely mechanical and I suppose most of us permanent civil servants are in reality only mechanical; I look upon those who on change of

---

[78] PRO: RG 29/1, p. 641.
[79] PRO: RG 29/5, pp. 469–70.
[80] Wright, *Treasury control of the Civil Service*, pp. 110–13.

government come in above us and rule us as the intelligent and intellectual, not requiring experience, but instantly understanding their new duties intuitively.[81]

It is difficult to imagine a permanent secretary of a small Whitehall department writing to a senior Cabinet minister in such terms today!

Graham was always ready to stand up for Farr, and to defend him from criticism. In the *Annual report* for 1856, for example, Farr had made disparaging comments regarding the powers of magistrates to decide whether or not inquests needed to be held.[82] He implied that magistrates avoided calling them in order to save on coroners' fees, with serious consequences for the protection of life and medical science. Graham dealt with complaints from magistrates in such a brusque manner, that he was roundly rebuked by Spencer Walpole, the Home Secretary. Graham excused himself as an old soldier inured to military habits, and indicated that in this particular case he was prepared to 'kiss the rod'.[83]

It should be no surprise, therefore, that many of the fiery passages of purple prose found in the *Annual reports* in this period are in fact in Graham's own contributions, rather than in those of Farr. Thus, in 1862 it was Graham's report, rather than Farr's Letter, which contained the classic discussion of the quality of the water supplied by the London water companies.[84] However, sections in Graham's reports are sometimes attributed to Farr, on the assumption that they had to have been written by the latter. But there is no direct proof for this, and it is not inconceivable that Graham deployed material prepared for him by his subordinate. The belief that Farr wrote Graham's reports may have originated with Major Greenwood's assertion to that effect in his work *The medical dictator*.[85] Greenwood, however, gave no evidence for this apart from their 'fruitiness', which he

[81] PRO: RG 29/2, insertion between pp. 131–3.
[82] GRO, *19th ARRG for 1856*, pp. 204–5.
[83] PRO: HO 45/6554.
[84] GRO, *23rd ARRG for 1860*, pp. xxxiii–xxxvii.
[85] Greenwood, *The medical dictator*, p. 102.

The expansion of the GRO's statistical functions

Table 3:3. Salaries in the GRO and other departments, derived from the Civil Service estimates for 1861/2

Grade	Average of other departments* (min–max.) £	GRO (min–max.) £
Junior clerks	98–240	80–150
Second class/assistant	263–390	160–240
1st class/senior	424–580	250–350
Principals/Heads of departments	534–780	400–600
Inspectors/surveyors	744–833	300–500
Secretaries/chief clerks	1050–1220	600–800

* The Office of Woods and Forests, Audit, Poor Law Board, Paymaster General, Board of Works.

Source: PRO: RG 29/1, p. 614.

felt must reflect Farr's authorship. As already noted, however, Graham was equally capable of being 'fruity' in his correspondence!

In his dealings with his staff, Graham was a Victorian paternalist. The clerks in the GRO appear to have been comparatively lowly paid, as can be deduced from the figures in Table 3:3, which are taken from a staff memorandum of October 1861.

As a result, the GRO's clerks were constantly petitioning the Treasury for an increase in salaries.[86] These pay claims were seldom forwarded to the Treasury without Graham's ringing endorsement. Fair pay, according to Graham, was not just a matter of efficiency but of justice.[87] Yet, despite such support, and his evident concern for his colleagues, Graham could be a ruthless manager when required to be so. He was quite prepared

---

[86] See, for example, PRO: RG 29/1, pp. 434–5.
[87] PRO: RG 29/2, p. 9.

to put what he saw as the good of the public service, and the views of his political masters, before the interests of his staff.

This can be seen clearly in the manner in which Graham saved the GRO from imminent collapse when he became Registrar General in 1842. Thomas Lister may have laid the foundations for the statistical functions of the GRO but neither he, nor William Farr, appears to have been able to administer a department of state. On his arrival Graham found that there was a backlog of nearly two years in the compilation of the registers of births, marriages and deaths, whilst the departmental accounts were in a 'very irregular and disorganised state'.[88] The latter took nearly two years to sort out, and had to be closed £930 in arrears because of the incomplete recording of the distribution of books and forms for registration purposes.[89] Graham also discovered that 12 clerks had been underpaid because of misunderstandings over probation periods, whilst 19 had not contributed enough to the superannuation fund because of another misunderstanding over the date at which contributions should be levied.[90] The Treasury was convinced that the work of the Office was so badly managed that the department must be grossly overstaffed.[91]

Still more seriously, Lister's lax administration had tempted some members of staff to abuse their positions. Graham found that Lister had made no checks on the accounts of the office keeper regarding postage, and that he had made payments to the latter without any covering invoices. Inquiries at the General Post Office revealed that on several occasions the amounts claimed by the office keeper for postage in fact exceeded the receipts for the entire postal 'beat' from Charing Cross to Temple Bar. Confronted with this the office keeper confessed to embezzlement on a large scale.[92] Graham's reaction to these problems was swift and decisive. The chief clerk and the office keeper were sacked, whilst the entire clerical compliment of the

---

[88] PRO: RG 29/5, p. 186; RG 29/1, p. 177.
[89] PRO: RG 29/1, p. 220.
[90] PRO: RG 29/1, pp. 229–30.
[91] PRO: RG 29/5, p. 181.
[92] PRO: RG 29/1, pp. 210–12; RG 29/5, p. 232.

Office was reduced from 87 to 60 by the introduction of task work, and by the simple expedient of introducing printed registers with pro-forma entries with gaps for individual-level information. This radically reduced the amount of transcription required.[93] Graham was thus able to catch up on the backlog of work in the Office whilst reducing the departmental estimates.[94] This endeared him to the Treasury who commended him on his running of the department.[95]

Graham subsequently used his reputation for efficiency to extract greater clerical resources and higher pay from the Treasury for the various sections of the GRO, including the Statistical Department. Thus in October 1864, one finds him opening a letter requesting salary increases for his staff with the words, 'I trust that their Lordships are aware that in the expenditure of public money I am not an advocate of extravagance; on the contrary, I contend that I have proved myself to be always studying economy . . .' He went on to calculate that his introduction of pro-formas had saved the public purse £36,000 in salaries since 1843.[96] This intervention led to the maximum salary of junior clerks being increased from £150 per annum to £260, with consequential increases further up the office scale.[97] Over the period 1843 to 1864, he was able to increase the number of clerks by nearly a third, thus returning to the staffing levels of the Lister period but at much higher levels of productivity (see Table 3:1). The provision of resources for Farr's work was thus dependent upon Graham's good relations with the holders of the Whitehall purse strings.

Graham also had to maintain discipline within the GRO, and in the registration service in general. This was a difficult matter since he did not control recruitment to the Office, which was by Treasury nomination. Graham described his predicament in a graphic account he gave to the 1860 Select Committee on Civil Service Appointments. On taking up his post in the GRO:

[93] PRO: RG 29/1, pp. 177, 183–4, 202, 212.
[94] PRO: RG 29/5, p. 228.
[95] PRO: RG 29/5, p. 219.
[96] PRO: RG 29/2, p. 7.
[97] PRO: RG 29/2, p. 7; RG 29/6, pp. 9–10.

I found persons there of very bad character; one person in that list, who was nominated by the Treasury had been an insolvent debtor; he had debts to the amount of £4,300, and assets to the amount of £130; and he had been imprisoned by the sentence of the court as a fraudulent debtor; that is one instance. Another man I sent with some money to the Bank of England, but he did not pay it in, and tried to impute the blame to one of the clerks in the Bank of England, as having received it; and there were several things of that sort. Then with regard to health, there was one man who I was forced to keep in a room by himself, as he was in such a state of health that he could not associate with the other clerks; he died soon afterwards.[98]

In addition, poor pay and insecure employment in the early years of the GRO led to problems of staff discipline. Junior clerks were constantly getting into financial difficulties. According to Graham, when supporting a staff petition for better wages in 1850, the clerks' pay was:

not sufficient to support them and their families respectably – they can only live and dress like mechanics, who comparatively are better paid. The consequence is that the civil service is degraded by the too frequent appearances of some of them in the Bankruptcy and Insolvent Debtors' Courts, their creditors are unpaid, their own characters are destroyed and their efficiency is much impaired, being overwhelmed with pecuniary difficulties they become depressed and spirit broken they lose all zeal and are unequal to the performance of their daily task; driven to borrow money from usurers at enormous interest, they neglect their public duties being occupied in devising plans to escape from or appease their merciless creditors, and finally are totally ruined, being compelled to quit the Civil Service.[99]

---

[98] *Report of the Select Committee on Civil Service Appointments*, p. 176.
[99] PRO: RG 29/1, p. 329.

The expansion of the GRO's statistical functions 75

This resulted in an unusually rapid turnover of staff, and those that remained had to work so much overtime that they often injured their health.[100] At least one transcriber of certificates is recorded as having been 'seized with mania' at his desk, and had to be removed to Bedlam.[101] It is evidence of Graham's authority, and perhaps of his ruthlessness, that despite such problems the great registers of births, marriages and deaths continued to be kept up to date, and statistical data generated on an annual basis.

Graham was equally conscientious in his supervision of the local registration service. The basic problem here was ensuring the proficiency of the thousands of registration official spread over the length and breadth of the entire country. Graham's solution was to appoint inspectors to examine the work of the local registrars, and to report on defects in their work. Four temporary inspectors toured the whole country in 1844 and again in 1845[102], as a result of which 'a considerable number of registrars who by personal examination and enquiry were proved to be incompetent and unworthy to be entrusted with such responsible duties . . .', were dismissed.[103] From 1846 two permanent inspectors were appointed with the intention of covering the whole country every four years.[104] Their work, which was essential to the efficient running of the service, was hardly enviable, since it was said to involve, 'constant exposure to all weather, travelling every day from village to village, frequently from farm to farm, living entirely at inns, often with the lowest class of accommodation and consequent liability to cold and rheumatism, seldom sleeping twice in the same bed.'[105]

One of Graham's greatest achievements was the administration of the decennial censuses from 1851 to 1871. It has been implied that because Farr was one of the census commissioners in these years he was in some sense responsible for their

---

[100] PRO: RG 29/1, pp. 250–1, 281.
[101] PRO: RG 29/1, p. 311.
[102] PRO: RG 29/5, pp. 229, 256.
[103] PRO: RG 29/1, p. 249.
[104] PRO: RG 29/1, p. 249; RG 29/5, p. 268.
[105] PRO: RG 29/1, p. 349.

organisation.[106] It is quite evident from the extant official correspondence, however, that the administrative and managerial ground work for the decennial enumerations was laid by Graham, whatever the intellectual input of Farr, or the latter's role in writing the reports. Certainly, the file of correspondence between the GRO and the Home Office over the taking of the 1851 census does not contain a single letter from Farr, all of them being written by Graham from the temporary Census Office in Craig Court.[107]

Until the passing of the 1920 Census Act established permanent powers for authorising the taking of the decennial enumeration, each census had to be authorised by a separate piece of legislation every ten years. Since the Victorian Census Acts[108] only sanctioned the immediate taking of the census, and the abstraction and reporting of the data amassed (which only took some three or four years), the census-taking apparatus had to be created from scratch on each occasion. The GRO did not have a permanent internal unit dedicated to the preliminary organisation of the census until 1904.[109] In addition, the Census Acts in the period whilst Graham was Registrar General were only passed some nine months before census night, thus leaving little time for the necessary work.[110] In 1851, moreover, there were two additional censuses, concerning attendance and accommodation at places of religious worship and in educational institutions.[111] The form of these surveys appears to have been Graham's work[112] but he may have rued their introduction

---

[106] Eyler, *Victorian social medicine*, p. 41.
[107] PRO: HO 45/3579. The issues covered in the correspondence included: the remuneration of local registrars; undertaking census abstraction by task work; arrangements and the form of the schedule for the religious census of that year; sanction for the forms and schedules used in the population census; whether or not a return of vaccination should be made at the time of the census; the printing of reports; the estimates of the cost of taking the census; procedures for enumerating the population aboard ship; and arrangements for taking the census in the Islands in the British Seas; etc., etc.
[108] For references, see: Higgs, *Making sense of the census*, p. 105.
[109] PRO: RG 29/3, p. 310; RG 29/7, p.184.
[110] For the dates of the Census Acts and of the taking of the census, see: Higgs, *Making sense of the census*, p. 105.
[111] Thompson, 'The religious census of 1851'; Coleman, 'The incidence of education in mid-century'.
[112] Thompson, 'The religious census of 1851', p. 242.

The expansion of the GRO's statistical functions 77

given the controversy that the 'religious census' generated. A survey of the provision of facilities for religious worship, was transformed by Nonconformists into a commentary on their relative strength vis-a-vis the Established Church.[113] Doubts expressed over the reliability of the data collected in this census, and that of schools, may explain why they were never repeated as part of the population census. Graham certainly opposed the involvement of the GRO in the taking of an educational census in 1868 on the setting up of the Board of Education.[114] It was the thought of the approaching enumeration of 1881 that forced Graham into retirement in 1879 at the advanced age of 78.[115]

On this evidence, it is something of an injustice to describe Graham as a weak man, or as the meek bursar to Farr's master of the college. Indeed, as Major Greenwood argued, there is a good case for saying that 'Graham made Farr possible'.[116] But also, to some extent Graham *was* Farr, in the sense that some of the supposed achievements of the latter were actually those of his superior. One man who was in a position to judge these matters correctly, William Farr himself, would certainly have agreed. As he noted in his last public address, a letter to *The Times* of January 1880:

> For more than 37 years I have had the pleasure to serve under Major Graham, and had constant cause to admire and respect the energy, ability, personal attention to details, and capacity for organisation which marked his successful control of civil registration. No one acquainted with his duties, or with the way in which they were performed by Major Graham, can either describe his post as a sinecure or refuse to recognise the value of the services of the late Registrar General, although of a distinctly different character to my statistical duties.[117]

---

[113] Ibid., pp. 242–47.
[114] PRO: RG 29/2, p. 76.
[115] PRO: RG 29/2, p. 282.
[116] Greenwood, *The medical dictator*, pp. 103–4.
[117] *The Times*, 20 January 1880, p. 8.

Whitehall's demand for medical statistics

Part of the burden of Graham's duties as Registrar General was his increasing correspondence with other parts of central government. Although the GRO's initial statistical role appears to have been directed at the initiation of personal or local action, once it had begun publishing data on mortality, it rapidly became a resource upon which other central government bodies wished to draw. By facilitating the supervisory activities of the central State vis-a-vis the localities, the GRO was helping to create that local-central polarity around which Victorian governance was constructed.[118] These constant external demands, fostered in part by the GRO's own propaganda campaign, also underpinned the Office's approaches to the Treasury for more staff resources. The GRO's development cannot be seen in isolation, therefore, or in terms of an internally generated intellectual project, but needs to be studied as part of the gradual evolution of the modern pro-active central state.

As early as March 1840 Lister was writing to the Treasury for permission for overtime working to handle a request from Parliament for a return of the number of people who died of smallpox in 1839. Lister estimated that it would take a clerk, working six hours daily, 50 or 60 days to examine the necessary 350,000 entries.[119] This was presumably linked to the passage of the 1840 Vaccination Act that was being debated in Parliament at this date. Similar demands were made on the resources of the GRO in 1871 by the parliamentary select committee investigating the workings of the Vaccination Acts.[120] Again, in May 1878 the Treasury directed an extract from an order of the House of Commons to the GRO requiring it to supply the numbers of deaths from hydrophobia during each of the 11 years from 1866 to 1877, in each county of England and Wales.[121] The collection of national statistics facilitated the emergence of Parliament as a

---

[118] For the centrality of this polarity to nineteenth century state information gathering see, Higgs, *The Information State in England,* pp. 64–9.
[119] PRO: RG 29/1, p. 70.
[120] PRO: RG 29/2, p. 128.
[121] PRO: RG 29/6, p. 197.

national, directing institution, as seen in the gradual eclipse of specific local acts of Parliament by general legislation applying to the whole country.[122]

Overtime working was not sufficient to cope with all such demands. At the beginning of 1846 Major Graham wrote to the Treasury indicating that he had been asked by the Home Secretary to improve the range of mortality statistics published in the *Annual report,* that would necessitate the addition of five extra staff to Farr's department. This was the result of an approach made to Sir James Graham by the British Association for the Advancement of Science. As the Registrar General explained to the Treasury, he had, 'consulted the most eminent actuaries in the Metropolis, the council of the Statistical Society of London, and others who have devoted their attention to the study of vital statistics', and had resolved that the causes of death should be abstracted and combined with ages at death, and the results published by counties, London and large towns. He also wanted these to be combined with occupational data from censuses to show rates of mortality and expectation of life in particular trades.[123]

The expanded resources of the Statistical Department were to be fully stretched by the demands placed upon it by the General Board of Health. The Board, set up under the 1848 Public Health Act, and with Edwin Chadwick as its secretary, appears to have regarded the GRO as a limitless resource upon which it could draw. Given Chadwick's role in the establishment of the Office, this is perhaps understandable. As already noted, local boards of health were to be set up under the Act if the death rate rose above 23 per 1,000 over the previous seven years. Consequently, the Board bombarded the GRO with requests for information on general death rates for specific towns and districts. In the course of 1852 alone, for example, the

---

[122] In 1758, for example, out of 71 acts of Parliament passed in that session (32 Geo. 2), 40 related to local matters. In the session of 1858, however, out of 110 acts only 14 were local or personal: HMSO, *Chronological table of the Statutes,* pp. 124–5, 408–12. For the expansion of Parliamentary information gathering, see Eastwood, 'Amplifying the province of the Legislature'.

[123] PRO: RG 29/1, p. 246. The Treasury agreed to this request in March of that year: PRO: RG 29/5, pp. 267–8.

Board asked for the general numbers of deaths per thousand over the previous seven years, plus the proportion of zymotic diseases, in Nottingham, Melton Mowbray, Wallingford, Hereford, Winchester, Gravesend, Huddersfield, Rochdale, Chorley, Marlborough, Tunstall, Kingston upon Thames, and Ipswich. The GRO also supplied unspecified returns for Mould, Trowbridge, Huddersfield and Sandwich in the same year.[124] In addition, the General Board asked the Office to prepare numerous ad hoc returns – on, for example, cholera and diarrhoea in 1850[125], on the number of deaths from consumption in Ipswich in 1852[126], on smallpox in 1853[127], and on cholera and diarrhoea in London in 1853, and nationally on a weekly basis in 1854.[128] This last set of returns was sent at the same time as the GRO was supplying the Home Office with a breakdown of the chief causes and ages of death in various occupations, which had in turn been suggested to Lord Palmerston by the General Board of Health.[129] The Treasury supplied Graham with six temporary clerks to meet this extra demand.[130]

Indeed, so voracious was the Board for information that it had its own staff based at the GRO to work on the registration data. On 10 June 1849, the Board's clerk, Mr Edwards, reported to it that he had just spent 13 weeks collecting information on the sanitary condition of the merchant marine, and was 'framing tables showing the sanitary state of the merchant marine as compared with that of troops on foreign service'.[131] In 1850 Edwards and another clerk were seeking permission to extract deaths on a daily basis for parts of the Metropolis in order to get an idea of seasonal variations in mortality.[132] Edwards still

---

[124] PRO: MH 13/260.
[125] PRO: MH 13/260, letter from the GRO of 18 July 1850. This return, or an earlier one requested in March 1850, was estimated by the GRO to require 2,400 hours of work at a cost of £480.
[126] PRO: MH 13/260, letter from the GRO of 4 November 1852.
[127] PRO: MH 13/260, letter from the GRO of 12 March 1853.
[128] PRO: MH 13/260, letters from the GRO of 1 August 1853 and 9 September 1854.
[129] PRO: T 1/5914A/20153: letter of 31 August 1854.
[130] PRO: RG 29/5, p. 378.
[131] PRO: MH 13/260.
[132] Ibid., letter of 5 November 1850.

appears to have been working in the GRO in 1855, along with three other Board clerks.[133]

Major Graham was prepared, if grudgingly, to accede to the Board's demands, although it was always duly billed for the work done. But in the summer of 1854 the Board was dissolved, and Chadwick ejected in a reaction to his over-centralising tendencies.[134] Although reconstituted, the Board never recovered its dynamism and was finally wound up in 1858. Almost immediately Graham moved to restrict the amount of work performed by the GRO for its sister department. By November 1854 he was writing to the Board turning down a recent request for extensive returns relating to cholera and diarrhoea. He suggested that Farr provide more modest tables, and as a consequence:

> Time would then be allowed for considering the whole question as to the best mode of furnishing from this Office from time to time the returns required by the General Board of Health. I have not objected to employing for this purpose occasionally one or two clerks after office hours – but upon a large scale and for furnishing voluminous returns, I very much object to the system, independent of its adoption being absolutely forbidden by the Treasury.[135]

A year later Graham was declining to do any work of this nature at all, although occasional returns continued to be made until the Spring of 1857.[136]

But the passing of Chadwick provided only a brief respite, since his demands were replaced almost immediately by those of Dr John Simon. Simon had been the first MOH for the City of London, and was appointed as medical officer of the reconstituted General Board of Health in October 1855. When the Board was abolished in 1858, he obtained a similar post under

---

[133] Ibid., letters of 9 May and 22 June 1855.
[134] Finer, *The life and times of Sir Edwin Chadwick*, pp. 453–74; Lewis, *Edwin Chadwick and the Public Health Movement*, pp. 358–75; Brundage, *England's 'Prussian minister'*, pp. 133–56.
[135] PRO: MH 13/260, letter of 13 November 1854.
[136] Ibid., letters of 27 October 1855 and 28 April 1857.

the aegis of the Privy Council. From this position Simon took up the running as the dynamic centre of public health administration in central government. Simon saw his function, much as did Farr, in the spreading of scientific knowledge which would create an 'educated local patriotism'. Local opinion, suitably leavened by the activities of local MOHs and wielding the local franchise, would force local authorities to undertake sanitary reform. Given the fate of Edwin Chadwick, he was more careful to balance central direction with local responsibility.[137]

The Simonian agenda had an almost immediate impact on the work of the GRO. In 1857 Simon got permission from Graham to have one of his own clerks, E. H. Greenhow, work through the GRO's mortality data to produce figures for deaths by various causes broken down by districts.[138] The resulting *Papers relating to the sanitary state of the people of England* went beyond the GRO's practice of only publishing total district mortality rates.[139] From 1858 Simon arranged for the GRO to provide him with notification of local epidemics of certain diseases and quarterly figures concerning vaccination.[140] Similarly, it was Simon who prevailed upon Parliament to require the GRO to produce the first *Decennial supplement* of mortality data covering the period 1851 to 1860.[141] The resulting shift from relatively short-term reportage to long-term analysis of statistical trends was to have important implications for the future. Finally, in the late 1860s Simon persuaded Graham to enlarge the quarterly returns to include details of the numbers and causes of deaths in subdistricts.[142] By asking for local mortality rates to be differentiated by causes, Simon was encouraging the GRO to move away from producing local data for actuarial purposes, and in order to enable citizens to make discriminating life choices, towards more specific medical

[137] Simon, *English sanitary institutions*, p. 474.
[138] Lambert, *Sir John Simon*, pp. 262–3.
[139] Greenhow, *Papers relating to the sanitary state of the people of England*.
[140] *First Report of the Medical Officer of the Privy Council*, pp. 23–5.
[141] *Supplement to the 25th annual report of the Registrar General;Fourth Report of the Medical Officer of the Privy Council*, pp. 35–6. A copy of the parliamentary order, dated 24 July 1863, can be found in PRO: T 1/6452A/17223.
[142] *First Report of the Royal Sanitary Commission*, QQ 1908–10; Second *Report of the Royal Sanitary Commission*, Q 9695.

research. In this, of course, he was pushing at an open door. The GRO was increasingly providing the means for the centre to supervise the localities, with all the implications this had for the power of the apparatus of the nation state.

Simon's biographer, Royston Lambert, has accused the GRO of weakness and tardiness in these matters in order to stress the creative role of his hero.[143] On the other hand, Simon Szreter has defended the GRO by pointing to its willingness to provide the data sought, and to the origins of some of Simon's innovations in the ideas of William Farr.[144] In a sense, both these approaches miss the real crux of the matter, which was not the relative intellectual prowess of Simon and the GRO, but the human resources that the latter had at its disposal to pursue both its own agenda and that of other departments. Simon's demands placed great strain on the GRO's limited statistical manpower. The production of the *Decennial supplement,* for example, required the employment of seven clerks on overtime.[145] Graham indicated to the Treasury that:

> were I to supply all the returns which Mr Simon, the medical officer attached to the Lords of the Privy Council persuades their lordships to consider essential, it would be necessary for me to ask the Lords of the Treasury to add permanently to this establishment one or two clerks possessing a certain degree of medical knowledge.[146]

Simon's requirements depended upon the provision of adequate human resources within the GRO but were also a means of obtaining them from the Treasury. Nor, should it be noted, was Simon doing anything new, since the pressure he placed on the GRO merely echoed that of Edwin Chadwick and the General Board of Health a decade earlier.

Even when the Office was attempting to meet the requirements of local government officials, such as MOHs, it increasingly

---

[143] Lambert, *Sir John Simon,* pp. 262–3, 320–1, 419.
[144] Szreter, 'Introduction: the GRO and the historians', pp. 410–11.
[145] PRO: RG 29/1, p. 632; RG 29/5, p. 467.
[146] PRO: RG 29/1, p. 603.

had to deal with nationally organised professional groups. The first MOH had been appointed for Leicester in 1846, to be followed by Liverpool in 1847, and the appointment of John Simon to the post in the City of London a year later. The 1855 Metropolis Management Act established MOH posts for London's vestries and districts, whilst the 1872 Public Health Act placed a duty to appoint them on all local health authorities.[147] The work of the MOHs covered all aspects of Victorian public health – sanitation, vaccination, quarantine, and the notification of infectious diseases. Diplomas in Public Health began to be offered in the universities following the legal requirement for properly qualified officers, whilst the Local Government Act of 1888 introduced statutory qualification requirements for public health appointees. By the end of the century MOHs had achieved security of tenure, releasing them from dependence on private medical practice.[148] The Metropolitan Association of Medical Officers of Health had been founded as early as 1856, and a number of provincial societies were established in the 1870s and 80s. These amalgamated in 1887 and began publishing their own journal, *Public Health,* in 1888. In 1891 the Society was incorporated under the Companies Acts as 'The Incorporated Society of Medical Officers of Health', with a membership of between 500 and 600.[149] It was this cadre of medical professionals who were the most avid consumers of the GRO's statistical output.[150]

As early as March 1867 the Registrar General was asking the Treasury's permission to employ clerks on overtime since:

> The medical officers of health in London and other large towns are urgent (sic) that a history should be *forthwith* published of the fatal causes of diarrhoea and cholera which occurred in the year 1866. The medical profession generally

---

[147] Wohl, *Endangered lives,* pp. 179–204; Hardy, *The epidemic streets;* Szreter, *Fertility, class and gender in Britain,* pp. 190–203.

[148] Hardy, 'Public health and the expert', pp. 130–7.

[149] Duffield, 'History of the Society of Medical Officers of Health', pp. 2–3. For the professionalisation of the MOH cadre, see: Porter, 'Stratification and its discontents'.

[150] Lewes, 'The GRO and the provinces'.

are anxious to be *promptly* furnished with this report, seeing that in the investigations made in connexion with this office strong evidence has been adduced that frequently the spread of cholera is to be attributed to bad water . . . . When their lordships look at the thousands which are spent under the direction of the Lords of the Privy Council in preparing histories, returns and reports of the cattle plague I think that they will not deem this proposal extravagant.[151]

The way in which the fortunes of the GRO came to be bound up in the wider politics of public health can also be seen from its involvement in the imposition of compulsory vaccination on a recalcitrant British public. The 1840 Vaccination Act had banned inoculation for smallpox and made vaccination against the disease more freely available. However, the 1853 Vaccination Extension Act went still further and made infant vaccination compulsory. This unprecedented imposition by the state on its citizens was energetically opposed by the Anti-Vaccination League, and similar organisations. Given the somewhat unpleasant, and dangerous, nature of early forms of vaccination, the anxiety of parents is understandable.[152] It was as late as 1898, with the Vaccination Act of that year, that the compulsory aspects of the system began to be removed.[153] Civil registration helped to enforce compulsion since under the 1853 Act doctors were to send duplicate certificates of vaccination to the local registrars, who were to make them available for searches. The GRO in London provided the books and forms that the local registrars needed for this work. In 1867 a new Vaccination Act laid down that the registrar was to deliver a notice to parents registering a birth requiring them to have their child vaccinated. Doctors or parents were still to send certificates of vaccination to the local registrar. However, the registrar was now to

---

[151] PRO: RG 29/2, pp. 64–6.

[152] MacLeod, 'Law, medicine and public opinion'; Williams, 'The implementation of compulsory health legislation'; Durbach, 'They might as well brand us'. This was an example of the popular opposition to state compulsion which can also be seen in campaigns against the Income Tax and Contagious Diseases Acts, and the conflicts over the Tichborne claimant's case: Joyce, *Visions of the people*, pp. 70–2.

[153] Lambert, 'A Victorian National Health Service'.

notify the local poor law guardians every six months if he had not received a certificate for a child whose birth they had registered so that the guardians could investigate. This somewhat *ad hoc* system was formalised by the 1871 Vaccination Act (1867) Amendment Act, according to which every registrar was to send a return of all births and deaths of children under 12 months to the local vaccination officer at least once a month.

Conflicts over compulsory vaccination played a key role in the passage of the 1874 Births and Deaths Registration Act. The Act radically altered the nature of the civil registration process by enforcing the certification of causes of death by a qualified medical practitioner, and by introducing fines on householders, or next of kin, for failing to register births and deaths. It has been suggested that these innovations were introduced to improve the quality of medical and demographic statistics, and this was certainly why they were recommended by the Royal Sanitary Commission in 1871.[154] However, the debates in Parliament reveal other, more urgent, considerations. Introducing the Registration Bill in May 1874, the President of the Local Government Board, George Sclater-Booth, noted that the necessity of an amendment to the law had been brought to the notice of the Government by the Report of the Sanitary Commission, the Committee on Vaccination in 1871, and the Committee on the Protection of Infant Life. He claimed that, 'Not only did births escape registration, but there was reason to believe that many burials took place of children born alive who were represented as still-born, and that was a defect which would be remedied by the present Bill.'[155] This reflected increasing concern respecting 'baby farming' and infanticide in the 1860s and early 1870s, which had led to the passing of the Infant Life Protection Act in 1872.[156]

Replying for the opposition, Lyon Playfair noted the need for accurate 'sanitary statistics', and he also wanted accurate death registration to prevent the concealment of murders. His

---

[154] Glass, *Numbering the people,* p. 181; Hardy, 'Death is the cure of all diseases'; *Second report of the Royal Sanitary Commission,* p. 58.
[155] *Hansard,* 3rd series, Vol. 219, May 14, 1874, col. 275.
[156] Sauer, 'Infanticide and abortion in nineteenth-century Britain'.

strongest language was reserved, however, for the threat to registration's role in the vaccination system:

> There is an active League, called the Anti-Vaccination League, which does not hesitate to induce parents to conceal births in order to prevent vaccination. The League will, unquestionably, be glad that this Bill contains no penalty for not giving information of births. Compare the infant mortality amongst children in Scotland, where registration is compulsory, with that in England, where it is voluntary, and you will be convinced how much efficient registration has to do with repression of that loathsome disease smallpox.[157]

By the time the Bill passed into law a clause introducing compulsory registration had been inserted. In the same session, Parliament was also discussing the 1871 Vaccination Act Amendment Act of 1874, which was passed to enforce the provisions of the earlier Act.[158]

The staff of the GRO saw both compulsory registration and medical certification in similar terms. Thus, as already noted, Farr was calling as early as the *Annual report* for 1864 for the compulsory registration of deaths under penalty on the grounds of the need to ensure that murders did not go undetected.[159] Similarly, a year before the passing of the new Registration Act, one finds Farr mounting a statistical defence of smallpox vaccination in his Letter attached to the *Annual report*.[160] When George Graham discussed the new legislation in his *Annual report* for 1875, he saw medical certification in terms of preventing crime rather than in producing better statistics. As he noted:

> Registered medical practitioners are now required under a penalty, to certify the causes of their patients' deaths, which are registered together with the names of the certifying practitioners. The number of uncertified deaths has already been

---

[157] *Hansard*, 3rd series, Vol. 219, May 14, 1874, cols 281–2.
[158] *Hansard*, 3rd series, Vol. 221, July 27, 1874, cols 836–7.
[159] GRO, *27th ARRG for 1864*, pp. 177–91.
[160] GRO, *34th ARRG for 1871*, pp. 219–21.

greatly diminished under the new statute, and the inquiries which the registrars now make when no medical certificate is produced cannot but tend to strengthen the protection to human life which registration supplies.[161]

As in the case of the original 1836 legislation, scientific research appears to have been a beneficiary of the 1874 Registration Act rather than the key factor in its passage through Parliament.

However, at the same time that the energies of the GRO were being diverted from its own agenda into that of other national state bodies, so some of its own ability to intervene via propaganda at a local level was being reduced. In 1858 the Treasury radically curtailed the GRO's practice of distributing free copies of its reports to learned societies, medical practitioners, reading rooms, mechanical institutes, and private individuals. This the Treasury felt was, 'a serious expenditure of public money, without any security being afforded that the documents would come into the hands of those who would duly appreciate them.' Whereas the GRO had previously received 5,000 *Annual reports*, 9,100 *Quarterly reports*, and no fewer that 65,100 *Weekly reports* per annum for free distribution, it was now to receive only, 100, 400 and 150 copies respectively. By 1876 the GRO was being allowed 350 copies of the *Annual report* for circulation.[162] In 1860 the entire print run of the *Annual report* appears to have been reduced to only 1,750 copies.[163] When, in the following year, Graham suggested to the Home Office that the Office's quarterly returns should be placed before Parliament and thus published as parliamentary papers, his proposal was rejected on the grounds that, 'the gratuitous distribution of papers is a very different matter from the circulation of useful knowledge . . .'.[164] The Office's ability to play an independent propaganda role in

---

[161] GRO, *38th ARRG for 1875*, p. ix.
[162] PRO: RG 29/2, pp. 551–7; RG 29/5, pp. 425–9; PRO: STAT 3/16 Outletters 1875–78, p. 268.
[163] *Return of the cost of reports and papers presented by command of Her Majesty . . . during the session of 1860*, p. 2.
[164] PRO: HO 45/6838 Registrar General's quarterly returns: refusal to lay before Parliament, 1859.

the localities was being reduced. This loss of independence was to accelerate in the years after the retirement of Graham and Farr at the end of 1879, and will be the subject of the next chapter.

4

Late Victorian medical statistics in an age of inertia

Evidence of institutional inertia

The last two decades of the nineteenth century have been seen by historians as a period of relative decline in the history of the GRO, sandwiched between the heroic age of Graham and Farr, and a new surge of scientific achievement in the early years of the twentieth century.[1] Certainly the *Annual reports of the Registrar General* in the period 1880 to 1900 were, on the whole, slighter texts than those of the High Victorian and Edwardian periods. As can be seen from Figure 3:2 above, the textual material they contained in the late Victorian period seldom exceeded 25 pages, which was only a third or a quarter of the average of the 1870s, and even less when compared to the *Annual reports* of the first decade of the twentieth century. Much the same could be said of the *Decennial supplements* in these years. Thus, that published under Farr's name in 1875, covering the period 1861 to 1870, contained 74 pages of text and 672 pages of tables. That produced ten years later by William Ogle contained only 61 pages of text and 436 pages of tables. John Tatham's first *Decennial supplement,* published in 1895 and covering the period 1881 to 1890, contained 57 pages of text but much more in the way of tables at 794 pages. But this was dwarfed by his second *Supplement* of 1907 that contained no less than 113 pages of text and 830 tabular pages.[2]

---

[1] This argument has been put forward most cogently by Szreter: 'The GRO and the public health movement', pp. 454–62.
[2] GRO, *Supplement to the 35th annual report of the Registrar General*; GRO, *Supplement to the 45th annual report of the Registrar*; GRO, *Supplement to the 55th annual report of the Registrar General*; GRO, *Supplement to the 65th annual report of the Registrar General*.

Quantity is not, of course, the same as quality, but the tone and content of the publications in the late nineteenth century also showed a certain falling off. As already noted, the Superintendent of Statistics' Letter specifically devoted to cause of death data was absorbed into the Registrar General's own report, in a much shortened form, in the 1880s and 1890s. The text of the *Annual reports* also lost much of the vivacity and pugnacity of the heyday of the Graham administration. The passages of purple prose, and the exhortations to local authorities to prevent unnecessary loss of life, were replaced by the rather dry recitation of statistical facts. The wording of some of the *Reports* in these years was indeed almost identical, with new dates and numbers merely being inserted in consecutive volumes. The GRO's *Quarterly reports* also show exactly the same retreat from overt polemicism in the years after 1880.

The reasons for this apparent loss of nerve in the late Victorian GRO are plainly complex. In a seminal article on the Office, Simon Szreter has defended the GRO's Statistical Superintendents in these years, William Ogle and John Tatham, as men of learning, intellectual curiosity and innovation.[3] Similarly, Anne Hardy points to Ogle's institution of the system of confidential enquiry between the GRO and certifying medical practitioners on the subject of ill-defined causes of death as a major advance in medical data gathering.[4] Ogle also moved beyond Farr's use of the 'Healthy Districts Death Rate' as the standard of mortality comparisons by introducing the concept of direct standardisation. Local death rates were now increasingly adjusted by basing their calculation on a standard population with a defined age and sex structure. Ogle began to use such standardisation in the *Decennial supplement* for 1871–1880, and it was introduced into the ARRG for 1901.[5]

---

[3] Szreter, 'The GRO and the public health movement', pp. 457–60. Ogle was statistical superintendent from 1880 to 1893, whilst Tatham held the post from 1893 to 1909.
[4] Hardy, 'Death is the end of all disease', p. 476.
[5] GRO, *The Registrar-general's Decennial Supplement, England and Wales 1921. Part III*, pp. xxxiii–xxxvii; Mooney, 'Professionalization in public health', pp. 60–1. See also Ogle, 'Proposal for the establishment and international use of a standard population', pp. 83–5.

This might be less useful for propaganda purposes but was a considerable improvement in scientific terms. Robert Woods has gone even further and has compared Ogle's intellectual achievements favourably to those of William Farr, drawing especial attention to Ogle's work on suicide upon which Emile Durkheim drew.[6] In addition, Tatham's list of appearances before parliamentary committees and commissions during his period as superintendent of statistics (1893–1909) hardly betokens a lack of due recognition.[7]

Rather than a product of deficiencies in scientific personnel, Szreter sees the malaise in the Office of this period as due to four main factors:

the GRO's provision of localised mortality statistics had come to be seen as scientifically outmoded because of the rise of the germ theory of disease causation, and of social darwinism and eugenics, which down graded the importance of the environmentalism inherent in the Victorian public health movement;

increased financial stringency, 'heralded with a Treasury minute of 1886, inaugurating a regime of inflexibility and refusal to countenance expansion in staff costs or improvements in pay';

the development of a 'bureaucratic and authoritarian ethos' with the rise of a professional audience for the GRO's output in the growing body of local medical officers of health (MOHs);

the differing personal styles of the successive Registrar Generals and their Statistical Superintendents – but only acting as 'triggers for the expression of the other underlying forces acting upon the GRO'.[8]

---

[6] Woods, 'Physician, heal thyself', p.3.

[7] He was an expert witness before the Inter-Departmental Committee on Gas Poisoning, the Royal Commission on Tuberculosis, the Select Committee on Death Certificates, the Royal Commission on the Feeble Minded, the Royal Commission on Arsenical Poisoning, and the Lords Select Committee on Infant Life Protection. He was also a member of the Inter-Departmental Committee on Physical Deterioration and of the Imperial Cancer Research Fund Committee: PRO: RG 29/3, p. 365.

[8] Szreter, 'The GRO and the Public Health Movement', pp. 453–7.

There is much truth in this explanatory model, although not always quite in the way that Szreter argues. The rest of this chapter will examine these, and other, possible factors for institutional retardation.

Causes of institutional inertia: bacteriology and eugenics

In the late Victorian period, medical scientists came to explain disease in terms of the effects of biological organisms, rather than of 'miasma', or of Justus von Liebig's theories of chemical poisoning.[9] According to Szreter, the development of bacteriology undermined, 'the scientific status of the GRO's principle method of investigation'. Comparative statistical inference, 'no longer held the high promise for scientific advances across a wide front that had seemed to be the case throughout most of Farr's era'.[10] Similarly, Linda Bryder has argued that germ theory fundamentally altered the nature of public health administration:

> Rather than focusing on sanitary engineering, public health now involved tracing and identifying the source of infectious diseases. Individuals thus traced would ideally be isolated and treated by vaccine therapy and their contacts vaccinated. Public health was thus becoming more concerned with the individual, and the laboratory was assuming a new importance.[11]

However, these arguments cannot be taken too far in the late Victorian period. Although there were important discoveries in bacteriology in the 1880s, such as that of the cholera bacillus by Koch in 1883, there were still disagreements into the 1890s over whether bacteria were stable species, the nature of their life-cycles, the conditions which favoured their multiplication, and how they

---

[9] Eyler, *Victorian social medicine*, pp. 97–108.
[10] Szreter, 'The GRO and the public health movement', pp. 456–7.
[11] Bryder, 'Public health research and the MRC', p. 59.

might be identified.¹² Similarly, although private medical laboratories were set up in the period before the First World War, and the possibility of establishing a public laboratory service was discussed, comparatively little was actually achieved in this sphere in Britain until the Second World War.¹³ Farr's own gradual acceptance of something approximating to the germ theory of disease did not dim his own faith in statistical epidemiology.¹⁴

Moreover, the argument that the spread of laboratory-based medicine undermined the statistical approach of the GRO is based on the assumption that the development of scientific methodologies necessarily takes place by replacement and exclusion rather than by accretion. There is no reason, however, why differing research techniques cannot complement each other in scientific practice. One should not confuse changes in research techniques with changes in scientific paradigms. When in 1914, for example, the Medical Research Council (MRC) set up its National Institute for Medical Research, it established a Statistical Department alongside its three departments of experimental research in bacteriology, biochemistry and pharmacology. Much of the early work of the new department's head, John Brownlee, lay in the very field of statistical epidemiology that Farr and his successors had championed. His first major task for the MRC was a statistical analysis of excess mortality in the boot and shoe industry, which extended the GRO's own work in the field of occupational health.¹⁵ Similarly, when the London School of Hygiene and Tropical Medicine was first mooted in 1921, it was envisaged that it would contain a Department of Epidemiology and Statistics, as well as others dealing with chemistry and biochemistry, immunology and bacteriology, applied physiology, medical zoology, and general sanitation and administration.¹⁶ If statistical epidemiology was

[12] Hamlin, 'Politics and germ theories in Victorian Britain, pp. 121–6.
[13] Bryder, 'Public health research and the MRC', pp. 59–82.
[14] Eyler, *Victorian social medicine*, pp. 104–8.
[15] PRO: FD 4/1 Special Investigation Committee upon the incidence of phthisis in relation to occupation: the boot and shoe industry, 1915.
[16] PRO: FD 5/11 Proposed establishment of the LSHTM: correspondence with MRC, 1921–1948: 'Draft memorandum as the basis of a communication from the Minister of Health to the Trustees of the Rockefeller Foundation'.

seen as an indispensable part of medical research at the very heart of British laboratory-based medicine, it is difficult to see why the development of bacteriology should have led to the eclipse of the GRO in the late Victorian period. Nor would it help to explain the apparent recovery of the GRO's vitality and innovativeness in the Edwardian period, as will be discussed shortly.[17]

The argument that eugenics and social darwinism blunted the drive of the GRO in the 1880s and 1890s is also difficult to evaluate. Szreter has elaborated his argument that the GRO found itself in an intellectual cul-de-sac in the late nineteenth century in his major work, *Fertility, class and gender in Britain 1860–1940*. Here he portrays the environmentalist credo of the public health movement as under sustained attack from the eugenic theories of men such as Francis Galton and Karl Pearson.[18] Eugenics took physical explanations of cultural and social phenomena to their ultimate conclusion by postulating measurable social phenomena as being due to heredity. Thus, unemployment and poverty were not caused by problems in the economic system but by failures of individual character bequeathed by inheritance. The Eugenics Society, for example, believed that 'pauperism is due to inherent defects which are hereditarily transmitted'. Since individual human 'vitality' and racial 'character' were determined by heredity, any attempt to preserve the lives of the 'unfit' via public health measures was counterproductive since this prevented the survival of the fittest from operating properly. The sickly were merely allowed to live to an age at which they could breed, and thus create more sickly individuals who would further undermine the health of the overall population. As Pearson told a meeting of doctors in 1912, 'Darwinism and medical progress are opposed forces.'[19]

To make matters worse, the poorer and supposedly less fit sections of society were perceived to be outbreeding their 'betters'. The result was that, as Karl Pearson repeatedly claimed,

---

[17] Szreter, 'The GRO and the public health movement in Britain', pp. 459–62; Szreter, *Fertility, class and gender in Britain, 1860–1940*, pp. 129–282.
[18] Szreter, *Fertility, class and gender in Britain*, pp. 93–107.
[19] Pearson, *Medical progress and eugenics*, p. 27.

25 per cent of the population were producing 50 per cent of the next generation. The racial mix of the British people, it was alleged, was therefore undergoing a rapid transformation, with the 'worst' stocks in the community increasing whilst the 'best' stocks were dying out. This process could signify nothing less than 'national degeneration'.[20] Francis Galton's specific contribution here was his argument that heredity determined such differences in intelligence between individuals. This fitted into the contemporary middle-class concept of social class as differentiation based upon inequalities of intelligence and skill.[21] Some eugenicists such as Caleb Saleeby were critical of the notion that the social distribution of 'intelligence' could be mapped upon the contours of the British class system but most assumed that the members of the middle classes were genetically more 'fit' that the 'lower' orders.[22]

Notions of the importance of innate personal characteristics in social life were not, of course, something new to the late Victorian period. Concern over the assumed moral failings and high fertility of the poor had been a general theme of British social policy going back to the New Poor Law and beyond.[23] In the mid-Victorian period Herbert Spencer believed that the progress of society was pushed forward by the survival of the fittest.[24] Similarly, Galton was writing to Darwin voicing his concern that the least 'fit' appeared to be outbreeding the middle classes in the 1860s.[25] The hereditary model of the causation underlying social phenomena was, however, given greater salience in the late Victorian period because of both micro-biology and macro-politics.

In the 1880s August Weismann was propounding the theory that the germ cells that controlled reproduction were distinct from somatic or body cells. This logically entailed the rejection of Lamarck's belief that characteristics acquired during life could be passed on to descendants. Weismann's ideas were important

---

[20] Searle, *Eugenics and politics in Britain*, pp. 26–7.
[21] Ibid., pp. 52–60.
[22] Soloway, *Demography and degeneration*, pp. 73–80.
[23] Mazumdar, *Eugenics, human genetics and human failings*, pp. 37–8.
[24] Abrams, *The origins of British sociology*, pp. 66–7.
[25] Jones, *Social Darwinism and English thought*, p.100.

to many eugenicists, who drew from them the further conclusion that environmental reforms could have only a very limited effect on individuals, who had the kind of characteristics they had by virtue of their germ plasma.[26] Prior to this date Lamarkian theories of evolution had allowed men such as William Farr to combine both a belief in the possibility of racial degeneration and of the efficacy of environmental improvement. Urban life was Farr's primary concern, and he believed that children who lived in cities with high mortality rates were likely to be crippled or weakened. Such disabilities were passed on to future generations, and so the results were cumulative.[27] The corollary of this concept, that environmental improvements would improve the racial stock, was an argument that many eugenicists of the late nineteenth century now believed could not be made consistently. Hence, Karl Pearson could oppose environmentalism to heredity in a wholly new manner – 'We have placed our money on Environment, when Heredity wins in a canter'.[28]

At the same time, the growing awareness of imperial crisis in the late Victorian period increasingly raised concerns in Britain over the future of the imperial race. The relative decline of British manpower resources in the late nineteenth century, as revealed by the GRO's own statistics, fed into debates respecting national efficiency. In the period 1870 to 1914 Britain's population increased more slowly than in almost every other major European state. When Germany was unified in 1871 its population was 41 million, 10 million more than that of the United Kingdom, and this gap had approximately doubled by 1914. Similarly, the USA was a country of 100 million by 1914, and expanding rapidly. This led to an emphasis on the Empire as a racial union of the 'English folk' scattered throughout the globe. Britain's relative economic decline was also marked, with the country's share of world trade in manufactured goods falling from 37.1 per cent to 25.4 per cent in the period 1883–1913,

---

[26] Searle, *Eugenics and politics in Britain*, p. 6.
[27] Eyler, *Victorian social medicine*, pp. 154–8.
[28] Searle, *Eugenics and politics in Britain*, p. 48.

whilst that of Germany rose from 17.3 to 23 per cent, and that of the USA from 3.4 to 11 per cent.[29]

The rise in the infantile mortality rate in the late Victorian period, from 146 per 1,000 in 1876 to 156 per 1,000 twenty one years later, also led to infant and child welfare becoming a burning political issue.[30] This perception of crisis manifested itself in numerous ways which often transcended party political boundaries.[31] British confidence was also undermined by the disasters of the Second Boer War (1899–1902). This imperial debacle can be explained in terms of British disorganisation and poor leadership but contemporaries blamed the poor physical condition of the troops enlisting in the army. General Sir John Frederick Maurice estimated that out of every five men who wished to enlist, there were only two remaining in the army as effective soldiers after two year's service.[32] This led to the creation of an Interdepartmental Committee on Physical Deterioration that took evidence on the subject in 1904. The committee was set up in the aftermath of the Boer War to investigate the possibility of a decline in the health and physique of the nation and certain social classes, and to suggest remedies.[33]

In the minds of the eugenicists, national efficiency and eugenics were inextricably entwined. As Karl Pearson put it in his Huxley Lecture at the Anthropological Institute in 1903, Britain was failing in its imperial struggle with the USA and Germany because of a lack of intelligence, and the only solution to this was:

> to alter the relative fertility of the good and the bad stocks in the community .... That remedy lies first in getting the intellectual section of our nation to realise that intelligence can be aided and be trained, but no training or education can <u>create</u> it. You must breed it ...[34]

---

[29] Searle, *The quest for national efficiency*, pp. 9–12, 34–53.
[30] Lewis, *The politics of motherhood;* Dwork, *War is good for babies.*
[31] Searle, *The quest for national efficiency*, pp. 54–97.
[32] Briggs, *Social thought and social action*, pp. 42–3.
[33] Searle, *The quest for national efficiency*, pp. 60–4.
[34] Pearson, 'On the inheritance of the mental and moral characters'.

According to F. C. S. Schiller, 'The nation which first subjects itself to a rational eugenical discipline is bound to inherit the earth.'[35] The life and health that the GRO saw as an individual's right had become an obligation to be healthy placed on the individual. If individuals could not, or would not, carry out their duties, extreme eugenicists envisaged them being removed from society, or prevented from passing on genes to the next generation. Reproduction was no longer to be a brute fact of life, as Malthus had conceived, but a process regulated by a nexus of rights and duties.

This was plainly an extremely imposing threat to the GRO's environmentalist project, and to the concept of the independent citizen capable of improvement and moral action upon which it was based. It is, however, somewhat difficult to explain the changing form of the *Annual reports* in terms of this intellectual challenge. What were the specific events in the late 1870s which took the wind out the GRO's sails, and what constellation of events in the late 1890s led to the expansion of its output in the new century? It is, of course, always difficult to prove that a specific event was 'caused' by changes in a broader intellectual framework. Thus, one would hardly expect the GRO to formally record that it was reducing the scale of its output, and changing its tone, because it doubted the scientific coherence of its own activities. Questions need to be asked, however, as to whether or not the timing and nature of the eugenic challenge was such that it can consistently explain such phenomena.

Although interest in eugenic ideas was a general feature of late Victorian society, it could be argued that the real impact of eugenics as an organised movement on social policy was not felt until the early twentieth century, especially in the aftermath of the Boer War and the deliberations of the Interdepartmental Committee on Physical Deterioration.[36] Although the biometric school of Karl Pearson and W. F. R. Weldon had emerged in the mid-1890s[37], it was only in 1901 that Pearson founded *Biometrika* as a journal for the discussion of biometric concepts,

---

[35] Searle, *Eugenics and politics in Britain,* pp. 34–6.
[36] *Report of the Interdepartmental Committee on Physical Deterioration.*
[37] Magnello, 'Karl Pearson's Gresham lectures'.

after the Evolutionary Committee of the Royal Society refused to publish the results of biometric research.[38] Similarly, Pearson and his colleagues only began to collect data on the link between fertility and social status *after* the publication of the Interdepartmental Committee's report in 1904, which had, in general, rejected eugenic ideas. Charles Spearman's first statistical experiments on the intelligence of children were not carried out until 1904, and the Eugenics Education Society only founded in 1907.[39] Similarly, the great debates between eugenicists and public health environmentalists such as Arthur Newsholme took place in the Edwardian period rather than in the last years of the nineteenth century.[40]

It should also be noted that the late Victorian GRO was quite capable of replying to the arguments of eugenicists in its own statistical terms. Indeed, the field of aggregate mortality statistics was one in which the GRO could, and did, score vital points for the environmentalist cause in that period. The first mention of eugenic arguments in the *Annual reports* was as early as that for 1879. The then Registrar General, Brydges Henniker, on noting an apparent rise in mortality rates amongst the middle aged and elderly, in contrast to steady improvements amongst younger age cohorts, commented that:

> it has been suggested that it is to sanitation that the increased death-rates of persons after a certain age should be referred. A vast number of children of permanently unsound constitution, are, it is said, now saved from death by sanitary interference. These grow up to adult life, and by their presence diminish the average healthiness of the adult classes, and so add to their death-rates.[41]

---

[38] Soloway, *Demography and degeneration*, p. 367, n. 35.

[39] Mazumdar, *Eugenics, human genetics and human failings*, pp. 41–51; Soloway, *Demography and degeneration*, pp 44–5. Richard Soloway also sees organised eugenics as essentially an early twentieth-century phenomenon: Soloway, *Demography and degeneration* pp. 1–17, 27. Donald A. Mackenzie also places the hey-day of the biometric school in the period 1900–1914: *Statistics in Britain*, p. 13.

[40] Eyler, *Sir Arthur Newsholme*, pp. 187–98. Pearson's attack on Newsholme's work on tuberculosis, for example, really began with the publication of the former's *A first study of the statistics of tuberculosis* in 1907.

[41] GRO, *42nd ARRG for 1879*, p. xxiii.

Henniker was able to dispose of this argument by noting that sanitation would save the lives of people at all ages, and that one would therefore expect rising mortality in all adult cohorts, not just those over 45, if the eugenic model was correct. Also the mortality rates for the elderly now seemed to be improving, which again did not square with eugenic arguments.[42] The relative decline in the mortality rates of all age cohorts over the age of 45 was something to which John Tatham was able to point to in the *Decennial supplements* for 1881–90 and 1891–1900.[43]

Lastly, the argument that the GRO's activities were curtailed in the late Victorian period because of a certain incoherence in it intellectual position in public debate, appears to run counter to Szreter's suggestion that it was increasingly retreating from general public debate to service the needs of medical professionals.[44] If its audience was predominantly professional MOHs, most of whom accepted the basic environmentalist credo, and were avid consumers of mortality data, why should a hypothetical intellectual inferiority to the eugenicists have any effect on the GRO's activities at all? As John Eyler has noted, a man such as Arthur Newsholme, whilst MOH for Brighton from 1888 to 1908, could still be 'a zealous and persistent student' of the *Annual reports* and the work of William Farr. Newsholme, a self-taught statistician, used the GRO's crude mortality rates to gauge local public health conditions, to study the epidemiology of diseases, and to construct local life tables. He was so enthusiastic about the GRO's statistical work that he was a candidate for the vacant post of Superintendent of Statistics in 1893, and told John Burns, the President of the Local Government Board (LGB), in 1907 that this was the only government post he really coveted.[45] The fact that William Ogle's successors, John Tatham and T. H. C. Stevenson (1909–1931), were both MOHs before joining the GRO, indicates the high esteem in which this professional group held the Office's statistical activities. To

---

[42] Ibid., pp. xxiii–xxiv.
[43] GRO, *Supplement to the 55th ARRG*, p. vi; GRO, *Supplement to the 65th ARRG*, p. xiii.
[44] Szreter, 'The GRO and the public health movement', p. 453; Szreter, *Fertility, class and gender in Britain*, pp. 190–203.
[45] Eyler, *Sir Arthur Newsholme*, pp. 27–51, 165–91, 219.

modern historians the work of the late Victorian GRO may appear to have lacked the relevance of earlier years but this does not mean that key contemporaries necessarily felt the same way.

Causes of institutional inertia: financial

Undoubtedly the late Victorian GRO faced serious financial difficulties that limited its ability to meet the increasing demands that were placed on it. It would be problematic, however, to see these as having their origins in a new financial regime suddenly introduced by the Treasury in 1886.[46] The GRO's staffing problems can be traced back to the late 1860s, and reflected changes in the overall grading structures of the Civil Service, as well as in the specific amount of money the Treasury was willing to spend on clerical resources within the Office. As can be seen from Figure 4:1, the clerical staffing levels in the GRO were fairly constant at about 77 or 78 in the period 1860 to 1874. There was then a modest rise to 86 clerks in about 1889, and then a return to previous staffing levels for the rest of the century. This was followed in the period from 1900 to 1920 by a rapid expansion in staff resources. During the period 1860 to 1900, however, the population of England and Wales increased by more than 50 per cent, from 20 million to 32 million. Each clerk in the GRO was thus dealing with a far heavier workload in 1900 than in the high Victorian period. The 1870s had seen some attempts to replace human with mechanical 'computers' but the pace of innovation appears to have slowed after about 1880.[47]

How far this pattern reflected any conscious government policy is difficult to determine. For much of the nineteenth century Parliament exercised little control over expenditure – it studied the estimates, voted supply and tied it to specified purposes with Acts of Appropriation – but there was no mech-

---

[46] Szreter's emphasis on the date 1886 possibly reflects his reading of Roy MacLeod's work on staffing levels in the Local Government Board: MacLeod, *Treasury control and social administration*, pp. 22–37. See, Szreter, *Fertility, class and gender in Britain*, p. 94, n. 58.

[47] Higgs, 'The General Register Office and the tabulation of data,', pp. 224–5.

Figure 4:1 Clerical staffing of the GRO

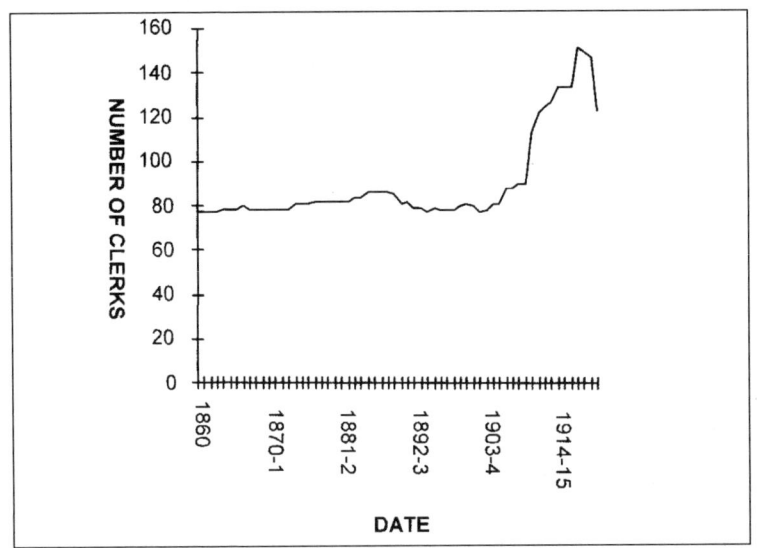

Sources: Civil Service estimates*

* This includes all clerical and senior staff of the GRO, plus transcribers, indexers, abstractors, copyists and sorters. Excludes industrial grades and typists. Prior to estimates for 1860 transcribers, indexers, abstractors, copyists and sorters not included. From 1889–90 copyists not included, number based therefore on the amount of money spent on copyists in 1888–89 (£2,740) divided by number of copyists (21) to give an average cost of a copyist (£130). This is then used to calculate the number of notional copyists in later years by dividing the amount spent on copyists by this figure.

anism for ensuring that the money was spent by departments in an appropriate manner. Parliament inched towards a solution of the problem in the course of the Victorian period. This process began after 1832 and was accelerated by the Exchequer and Audit Departments Act of 1866 which led to the absorption of the Exchequer into a greatly strengthened Audit Office, with the Comptroller and Auditor General submitting findings to the Public Accounts Committee. The Treasury was encouraged by Chancellors of the Exchequer such as William Gladstone to enforce the strictest standards of financial propriety on departments.[48]

[48] Roseveare, *The Treasury*, pp. 135–41.

How successful 'Treasury control' was in the Victorian period in controlling the size of the Civil Service is, however, somewhat debatable. As Maurice Wright has noted, the Treasury relied wholly upon the good will of the departments for the fulfilment of conditions written into increases in establishments. Apart from the memory of permanent officials, the Treasury had no administrative machinery for checking the filling of vacancies, the expiration of temporary appointments, the abolition of offices, and other such conditions.[49] Indeed, the development of Treasury control in the late nineteenth century can be seen as predicated upon the growth of the Civil Service, since new or expanding departments such as the Board of Trade and Colonial Office were inevitably forced to cross the Treasury's path in search of larger complements and salaries. Sir Reginald Welby, the Treasury's permanent secretary from 1885 to 1894, informed the Select Committee on National Expenditure in 1902 that since the 1880s, 'the wind was in the sails of the spending departments', supported by a Parliament less addicted to thrift.[50] Although there are problems, of course, in defining what exactly constituted a civil servant, the general trend, as shown in Table 4:1, appears to have been upward. In the mid-Victorian period there was, moreover, little control over hours worked and leave entitlements in individual departments.[51] The history of staffing levels in the GRO in the Victorian period should not, therefore, be read off any simplistic model of increasing or decreasing Treasury stringency.

A sea-change in the attitude of the Treasury to the GRO does, however, appear to have taken place in December 1869, when a Treasury minute was circulated suggesting the replacement of permanent civil servants performing 'mechanical' work with temporary writers and boys. The latter were not established and so could be dismissed without costly pensions.[52] The division of the Civil Service into intellectual/mechanical grades came to a head in 1871 when an order in council was issued

---

[49] Wright, *Treasury control of the Civil Service*, p. 174.
[50] Roseveare, *The Treasury*, pp. 148, 199–200.
[51] Wright, *Treasury control of the Civil Service*, pp. 296–7.
[52] PRO: RG 29/2, pp. 107–9; T 243/4, p. 145.

Table 4:1: Size of the Civil Service

Year	Number of civil servants	
1832	21,305	
1841	16,750	(excl. clerks, messengers, etc.)
1851	39,147	
1861	31,943	
1871	53,874	(incl. some workmen)
1881	50,859	(excl. Telegraph and Telephone Service)
1891	79,241	
1901	116,413	(incl. GPO)
1911	172,352	(incl. Telegraph and Telephone Service)
1914	280,900	(incl. Scotland and Ireland)

Source: H Finer, *The theory and practice of modern government*, vol. II (1932), pp. 1294–5

which strictly delineated the distinction between established and non-established civil servants – the latter were now to be called 'writers' and be paid at a uniform 10d an hour, whilst their privileges of sick pay and leave were to be abolished.[53]

As already noted, George Graham at the GRO was one of the few heads of departments who objected to the introduction of the distinction between intellectual and mechanical work in the Civil Service.[54] He had sat on the 1860 Treasury Committee on the Employment of Supplementary (i.e. unestablished) Clerks, which had recommended placing writers in a Central Copying Office rather than having them in the same department as established clerks. The committee's reasons for this were essentially to do with the internal morale of departments. In their report the Committee stated that:

> we have come to the conclusion that the maintenance of two distinct orders of clerks in the same office must, in the great majority of cases, lead to difficulties and embarrassments which cannot be easily or permanently surmounted. Every improvement in the supplementary clerks tends to render the distinction between them and the established clerks more

---

[53] Cohen, *The growth of the British Civil Service 1780–1939*, pp. 87–90, 127.
[54] The only other department that objected formally was the highly independent Foreign Office: Wright, *Treasury control of the Civil Service*, p. 184.

odious and untenable, so long as they are equally borne on the books of the same department; and every privilege which is conceded to the supplementary clerks in one office leads to demands for corresponding privileges in other offices; and increases the temptation to make unsuitable appointments from too high a class.[55]

Graham was concerned above all about the dilution of experienced staff within the GRO.[56] The concerns raised by Graham in this matter were somewhat reminiscent of contemporary objections to the creation of 'core' and 'peripheral' workers through corporate restructuring in modern industry.[57]

Graham appears to have ignored the Treasury minute of 1869, and the Treasury returned to the subject in August 1870.[58] The Registrar General begged leave to put off any restructuring until after work on the 1871 census was completed, if not 'the census will be a failure, and this department disgraced and £120,000 wasted.'[59] The Treasury agreed but insisted that in the meantime any vacancies were to be filled with unestablished writers obtained from the Civil Service Commission.[60] Over the next few years Graham expressed his unhappiness with this new regime[61], and was unwilling to replace staff with writers since 'as soon as they have learnt their duties, they depart hoping to obtain elsewhere better paid employment.'[62] Eventually, in March 1874, Graham forwarded a proposed new staffing structure for the Office in which 6 superintendents, 14 senior clerks and 32 assistant clerks were to be replaced by 5 superintendents, 10 senior clerks, 18 assistant clerks and 22 writers. This increased the number of staff in the GRO whilst reducing the salary bill by over £2,000 per annum.[63] By 1878 Graham appears to have

---

[55] PRO: T1/6250A/9059, p. 2.
[56] PRO: RG 29/2, pp. 107–9.
[57] Sampson, *Company man*.
[58] PRO: RG 29/6, p. 65.
[59] PRO: RG 29/2, p. 119.
[60] PRO: RG 29/6, p. 66.
[61] PRO: RG 29/2, pp. 131–3, 138–9, 151–2.
[62] Ibid., p. 155.
[63] Ibid., pp. 175–85. The Treasury agreed to this in August 1875: PRO: RG 29/6, p. 122.

become resigned to the use of writers, and was retiring established clerks in order to replace them with unestablished staff.[64]

In the late 1870s Graham was also being called upon by the Treasury to implement the recommendations of the Playfair Commission. This had suggested that the established clerks in departments should be divided into higher and lower divisions, with an increased proportion of business carried out by the latter on salaries ranging from £80 to £200 per annum.[65] The Treasury told Graham in 1875 that it was unwilling to allow any new established clerks to be appointed except to the lower division.[66] However, the Treasury took a still harsher line with Graham's successor, Brydges Henniker, by announcing in May 1881 that no new appointments to the GRO's establishment should be made until the whole redundant class of assistant clerks, 16 in number, had been replaced on retirement by lower division clerks.[67] This led to an explosion of staff resentment in 1885 with both the lower division and assistant clerks petitioning the Treasury for redress, claiming that the dilution of the senior staffing of the office reduced the chances of promotion.[68] Henniker also complained that the reduction in the number of senior and experienced assistant clerks created problems of supervision, whilst the clerks themselves noted that 'any permanent and satisfactory distribution of the work of the office amongst the different classes, with due regard to the nature of the work and pay of the clerks is practically impossible . . .'.[69] The Treasury, somewhat callously, suggested that the problem of 'stagnation in promotion' would be solved by retiring the GRO's secretary and two superintendents, 'who are all over 64 years of age, and have long earned their pensions'.[70]

In October 1887 the Treasury added to the GRO's problems by drawing the attention of heads of departments to the short

---

[64] PRO: RG 29/2, p. 270.
[65] *First report of the Civil Service Inquiry (Playfair) Commission*, pp. 14–15.
[66] PRO: RG 29/2, pp. 141–2.
[67] Ibid., pp. 372–3. This process only appears to have been completed in 1895: PRO: RG 29/3, p. 168.
[68] PRO: RG 29/2, pp. 356–66.
[69] Ibid., pp. 372–3, enclosure between pp. 375–6.
[70] PRO: RG 29/7, pp. 1–2.

six hour day worked by many lower division clerks, and suggested that they might think of moving to a seven hour day, with consequential staff reductions, in return for magnanimously recognising a half hour lunch break and a half day holiday on alternate Saturdays.[71] No immediate action appears to have been taken but in March 1889 Henniker suggested bringing in seven hour working, and giving lower division clerks an extra £15 as a consequence. He suggested that vacancies in the lower division should not be filled, and that an extra lower division clerk and a copyists could be dispensed with.[72] The Treasury was implacable in its reply, insisting that:

> My Lords are unable to understand why 19 clerks of the lower division at seven hours daily should be incapable of performing the duties of 22 clerks working only 6 hours, since the total in hours of the daily attendance of the 19 would be 133 hours, & of the 22, 132. My Lords find it necessary to insist on this reduction, in order to guard against the adoption of the seven hour system merely as a means of increasing the salaries of the lower division clerks without corresponding increase in work.[73]

The following year the Treasury was demanding that after the pressure of work consequent upon the taking of the 1891 census had ceased, that the permanent complement of the Office should be reduced by a seventh to take into account the length of the new working day.[74] The Treasury appeared unaware, or unconcerned, that the large amount of over-time working in the GRO meant that there would be no automatic proportional increase in staff output consequent upon an increase in the official daily attendance.

Staffing resources were plainly tight in the GRO in the late Victorian period but it should be noted that the resulting problems were not necessarily distributed equally throughout the

[71] Ibid., pp. 21–3.
[72] PRO: RG 29/3, p. 62.
[73] PRO: RG 29/7, p. 37.
[74] Ibid., p. 56.

Office. As can be seen from Table 3:2, the Record Department appears to have borne the brunt of the cuts, whilst the Statistical Department continued to expand between 1866 and 1895, from 16 to 19 clerks.[75] This increase of 18 per cent hardly kept pace with the increase in population, in excess of 40 per cent in this period, but must have brought some respite when coupled with the introduction of mechanical aids to computation and, possibly, the lengthening of hours worked.[76] Thus, the timing and nature of the changes in the GRO's publishing endeavours cannot be explained solely by Treasury intransigence. The Office's staffing difficulties began in the 1870s rather than the 1880s, and the Statistical Department may have escaped their worst effects in the later decades of the Victorian period. Overall, however, there was a gradual downgrading of the relative pay and conditions of the GRO's clerks, although it is difficult to establish how far this affected levels of skill and supervision.

Causes of institutional inertia: diversification of activities

But if the manpower resources of the Statistical Department increased in the late Victorian period, so too did the range of its activities. This process of diversification reflected both internal initiatives and external impositions. As already noted, when William Ogle became Superintendent of Statistics in 1880 he initiated a procedure for returning death certificates with vague causes of death to the responsible medical practitioners. By 1895 this was the first duty of one of the Office's statistical abstractors.[77] Another new activity for the Statistical Department was the redistribution of deaths in metropolitan hospitals back to the place of residence of the deceased, which will be discussed in more detail below.

---

[75] Even a careful scholar such as John Eyler appears on occasion to confound the staffing of the GRO as a whole with that of its Statistical Branch: Eyler, *Victorian social medicine*, p. 46.
[76] Higgs, 'The General Register Office and the tabulation of data', pp. 224–5.
[77] PRO: T 1/8954A/13154.

The standardisation of administrative boundaries, which the work of the GRO had helped to inaugurate led to an increased workload for clerks attempting to keep the Office's own records up to date. The failure to do so would mean increasing inaccuracies in mortality rates for named administrative areas given in the GRO's publications. The Office's attempts to create consistent time series for the population and mortality of named administrative units highlighted the complexities and local idiosyncrasies of English administrative structures. There were, for example, differing kinds of parish in the early nineteenth century – ecclesiastical, civil and administrative; differing amalgamations of parishes for administrative purposes in differing parts of the country – hundreds, wapentakes, rapes and lathes; 'detached parts' of one district imbedded in another; and some places which were, in theory, not in any recognised administrative units at all. These 'extra-parochial places', and larger liberties, enjoyed freedom from the usual burdens of local government, such as maintaining the poor, the militia laws, and repairing the highways. The creation of registration districts based on Poor Law unions under the 1836 Registration Act further complicated matters, since the registration counties made up of such districts were seldom the same as the ancient counties of England and Wales.[78] Such idiosyncrasies undermined the possibility of providing meaningful statistical comparisons of local conditions.

The Office's revelation of the country's chaotic administrative arrangements, uncertain administrative boundaries, and local peculiarities, provided the starting point for a gradual reorganisation of administrative arrangements. The guiding principle of local government organisation shifted over the nineteenth century from local rights and liberties, to the provision of nationally mandated services in standardised units.[79] In 1857, for example, a start was made on bringing the whole country within the normal systems of local administration with the passing of the Extra-Parochial Places Act. This decreed that places

---

[78] For a description of the multi-various administrative units recorded in the nineteenth-century censuses see, Higgs, *Making sense of the census*, pp. 127–133.

[79] Lawton, 'Census data for urban areas'.

named as extra-parochial in the *1851 census reports* were to be deemed parishes for the purposes of Poor Law administration, and were to appoint overseers of the poor. By the early years of the twentieth century the GRO's Statistical Department had a small unit solely concerned with establishing local boundaries for census purposes.[80] The provision of data on local populations was also utilised for the adjustment of the boundaries of parliamentary constituencies, thus underpinning concepts of fair and equal representation in the British polity. In this piecemeal manner, administrative and political arrangements in England were gradually standardised and unified in the same manner as they had been in nineteenth-century France.[81]

This process accelerated in the last 25 years of the nineteenth century. As Henniker explained to the Treasury in 1884:

> The Divided Parishes and Poor Law Amendment Act of 1876, as amended and extended by the Poor Law Act 1879, has resulted in an annually increasing number of Local Government Board Orders dealing with isolated and detached parts of divided parishes, altering the limits and incidence as regards registration, of a considerable number of districts . . . the Divided Parishes Act of 1882 has given rise to much correspondence and work has not been completed; a great strain has been thrown on the working power of one Branch of the Office.

In the same year Gladstone requested via the LGB that the GRO provide him with information on electoral boundaries in order to prepare legislation on parliamentary reform.[82]

This branch was, of course, the Statistical Department, where by 1898 a copyist and two or three boy clerks were occupied in noting county council orders regarding boundary changes; Local Government Board (LGB) orders respecting the boundaries of administrative counties or municipal boroughs; Acts of

---

[80] PRO: RG 29/3, pp. 195–7.
[81] For the situation in France see, Perrot and Woolf, *State and statistics in France*, p. 89.
[82] PRO: RG 29/2, p. 335; PRO: RG 29/2, p. 340.

Parliament affecting parliamentary constituencies; and the annual reports of the ecclesiastical commissioners. Since 1894 they had dealt with 910 LGB orders; the creation of 9,706 parishes, 89 urban districts, and 161 ecclesiastical districts; 797 changes to registration districts; 1,603 changes to parochial boundaries; and 211 changes to urban districts.[83]

Such activities can be seen in terms of a deep-seated drive on the part of the State to homogenise and control society via the creation of an all-encompassing administrative grid to allow it to pin down and control its subjects.[84] However, it is equally likely that what was being controlled was not society but local government, and that the standardisation of administrative space reflected the shifting relations between local and central government. The standardisation of administrative boundaries facilitated the monitoring of the former by the latter, rather than the reduction of all citizens to the mere objects of government.[85]

Other legislation also placed increased burdens on the statistical resources of the Office in this period. In 1894 Parliament passed the London Equalisation of Rates Act, which enabled rates to be standardised across the Metropolis via the creation of an Equalisation Fund. This Act also authorised the taking of a census for the purpose of ascertaining the numbers of persons present within each parish in the administrative county of London on the night of 29 March, 1896. The London County Council paid the GRO £150 for four second division clerks and 10 boy clerks or copyists to undertake the work. These were to be temporary postings from other departments but members of the Statistical Department had to supervise them.[86]

The debate on the introduction of national forms of welfare entitlements also began to affect the work of the GRO. In 1896 a select committee had been set up to consider, 'any schemes that may be submitted to them for encouraging the industrial population, by State aid or other wise, to make provision for old

---

[83] PRO: RG 29/3, pp. 195–7.
[84] Curtis, *The politics of population*; Hannah, *Governmentality and the mastery of territory*.
[85] Higgs, *The Information State in England*, pp. 64–9.
[86] PRO: RG 19/42 1895–1914 Census of London 1896; proposed censuses in 1906 and 1916: correspondence.

age ...'. Although this body, which reported in 1898[87], came to no firm conclusions, a small departmental committee was established to cost the various proposals it considered. The GRO was authorised by the Treasury in September 1899 to employ as many clerks as necessary 'on tabulating the information with regard to the numbers of persons of 65 years of age ...' for the purposes of this body.[88]

Increasing institutional subordination

As already noted, the second half of the nineteenth century saw a gradual professionalisation of the audience for the GRO's publications in terms of the rise of the nationally organised cadre of MOHs. But the most crucial external factor affecting the GRO's statistical activities was most probably a changing relationship with its parent department. Whilst the GRO was under the nominal control of the Home Office, the Registrar General had considerable freedom of action. This freedom began to be curtailed after 1871 when the Office's minister in Cabinet became the President of the LGB. Whereas before 1871 the work of the GRO was peripheral to the interests of its parent department, this was no longer the case when the Office found itself under a body directly responsible for local government and public health. Increasingly, the GRO was forced to concentrate more on servicing the needs of the Board in Whitehall, and less on its own project of local propaganda. Under Graham this reorientation was limited but became more pronounced after his retirement at the end of 1879.

In general terms, the establishment of the LGB, a fusion of the Poor Law Board, the medical wing of the Privy Council under John Simon, and the GRO, has been seen as a retrograde event in the fields of public health and social policy. The institution was, it is argued, dominated by desiccated bureaucrats from the old Poor Law Board who radically curtailed the initiative of the medical staff. Simon resigned from his post as the

---

[87] *Report of the Departmental Committee on Old Age Pensions.*
[88] PRO: RG 29/3, p. 233; RG 29/7, p. 141.

LGB's medical officer in 1876, ostensibly because of the refusal of its permanent secretary, John Lambert, to request Treasury sanction for three additional medical inspectors.[89] But behind this lay a difference of opinion over the role of the medical inspectorate. Rather than initiating action, Lambert believed that medical experts should be essentially sub-ordinate to the lay secretariat of the LGB at the centre. They should be peripheral and consultative, without powers of initiation or decision, and should exist mainly to provide advice when it was asked for. In the words of Simon's biographer, 'The records make it perfectly obvious that this decline was due to the dominance of the Poor Law mind, to a shrivelling of attitude, a cooling of sanitary ardour and a reluctance to interfere at the centre, not to sanitary impeccability among the localities'.[90]

Christine Bellamy, however, has presented a rather more nuanced picture of the Board, and the constraints upon its activities. The LGB was certainly bureaucratic and reactive, working mainly through casework generated by the exercise of its statutory duties. Precedent and policy evolved through the processing of applications submitted by local authorities.[91] The Board was also extremely hierarchical, and independent action on the part of its inspectorates was deprecated.[92] There were, however, certain structural reasons for the LGB's policy of not interfering too much with the localities. The LGB was placed in the invidious position of being both the sponsor of local government activity and the guardian of its financial probity. It was conscious of the need to placate local ratepayers, the Treasury, and the money markets, by keeping the rates and levels of local authority indebtedness as low as possible.[93] The LGB also shared the common Victorian conviction that local authorities and elites had a legitimate sphere of independent action, although they

---

[89] MacLeod, 'The frustration of state medicine', p. 17.
[90] Lambert, *Sir John Simon*, p. 541. The belief that the rigid bureaucratisation of the LGB's lay secretariat undermined its medical wing is widespread: Brand, *Doctors and the state*, pp. 22–36; Searle, *The quest for national efficiency*, p. 22; MacLeod, 'Introduction', pp. 15–16; Davidson, *Whitehall and the labour problem*, pp. 190–1.
[91] Bellamy, *Administering central-local relations*, p. 137.
[92] Ibid., pp. 125–6.
[93] Ibid., pp. 100, 259.

might be educated in a shared national ethos.[94] This model of central/local relations was based on practical considerations as well as ideological commitment. The growing volume, complexity and detail of the legislation it had to administer contributed to a profound sense of overload at the Board, especially given its staffing problems in the late nineteenth century.

All this meant that the LGB was primarily concerned to develop and protect its precarious influence by establishing a working partnership with local political elites. Its *modus operandi* was diplomacy rather than propaganda or compulsion, hence the Board's preference for general inspectors acting as 'ambassadors' rather than technical officers in the role of public prosecutors.[95] Even the Board's outstanding chief medical officers, John Simon and Arthur Newsholme, believed that the central state should act through local authorities in matters of health and welfare.[96] The fate of Edwin Chadwick a generation earlier had plainly left a legacy at the centre of Whitehall. This increasing cautiousness also needs to be seen within the context of a general decline in initiative in the Civil Service in the decades after 1870. This, it has been argued, was due to its increasing closure to recruitment from other spheres of activity, its subordination to an often indecisive Parliament, and a lack of consensus on policy issues.[97]

But this desire to limit its intervention in local government did not mean that the LGB was quiescent in its relations with the GRO. Like the General Board of Health and the Privy Council before it, the LGB was eager to use the GRO to monitor local compliance with national norms, but it now had the constitutional authority to impose its will. As soon as the GRO passed under the Board in August 1871 the amount and significance of the correspondence passing between the GRO and its parent department appears to have increased substantially. It is necessary to say 'appears' because of the destruction of so many

---

[94] Ibid., passim.
[95] Ibid., pp. 10–14, 116–26; MacLeod, 'Introduction', pp. 16–17.
[96] Eyler, *Sir Arthur Newsholme*, pp. xv, 207–19, 232–3.
[97] Davidson and Lowe, 'Bureaucracy and innovation in British welfare policy, pp. 44–52.

of the original Home Office files but from what material does survive the LGB certainly seemed more concerned with policy than the GRO's first parent department. Gradually a closer working relationships between the GRO and the Board developed, as the LGB began to involve itself in the work of the local registration system. In 1872, for example, the Board began paying for some of the activities of the local registrars, and in the following year it supported the GRO in making representations to the Treasury respecting increased remuneration for them in any new Registration Act.[98] By 1877 the Board was issuing its own orders directly to the registration service, the first time any authority other than the Registrar General had done so in the 40 years of its existence.[99] Groups such as the Registrar's Association and anti-vaccination campaigners began lobbying the Board with respect to the GRO's activities, whilst individual registrars complained directly about their pay and conditions. The anti-vaccination campaigners were complaining about Farr's statistical defence of vaccination against smallpox published in 1873 in the *34th ARRG for 1871*.[100]

This new relationship was deepened by the appointment of Sir Brydges Henniker to the post of Registrar General in 1880. Henniker was born in 1835, the son of Sir Augustus Brydges Henniker and a cousin of Lord Henniker, and succeeded to his father's baronetcy in 1849. Educated at Eton, he served in the 68th Foot, the Horse Guards, and was a captain in the West Essex Yeomanry. More crucially for the argument here, he was also private secretary to George Sclater-Booth, President of the LGB, at the time of his appointment to the headship of the GRO.[101]. Henniker's subsequent liaison with the LGB was so close that by 1894 he was asking the Treasury, 'that this office should be placed in telephonic communication with the LGB, with which department I am in constant and intimate official

---

[98] PRO: MH 19/192, letters of 11 April 1872 and 6 May 1873.
[99] PRO: MH 19/193, letter of 27 February 1877.
[100] PRO: MH 19/192, letters of 9 December 1872 and 12 December 1873; PRO: MH/193, letter of 27 June 1878; GRO, *34th ARRG for 1871*, pp. 219–21
[101] *The Times*, 8 January 1880, p.11, col. a; Nissel, *People count*, p. 147.

correspondence'.[102] The *Lancet* saw Henniker's appointment as 'evidence of the singular fatality which appears to have attended the ill-starred connection between Poor Law and public health administration . . . [and of the] . . . repressive policy adopted towards the sanitary and medical department by the central authorities of Poor Law administration.'[103] Henniker certainly started to rein in the GRO's publishing activities almost immediately. His first *Annual report*, covering 1878, but dated 31 March 1880, contained Farr's last Letter to the Registrar General, a comparatively short piece of 38 pages. The next *Annual report* abandoned the separate Letter altogether, and the Registrar General's own commentary began a contraction in size which was to continue into the 1890s.[104]

The LGB also began to exercise direct influence over the work of the GRO's Statistical Department. In March 1887, for example, Henniker informed the Treasury that, 'I am directed to obtain on behalf of the LGB as soon as possible returns as to the employment of the working classes in certain districts of the Metropolis, and to tabulate the results of the enquiry in this Office.'[105] The actual survey involved extensive enquiries made amongst the working-class populations of St George in the East, Battersea, Hackney and Deptford by 'enumerators' selected in the same manner as those for the census.[106] The work generated over 30,000 individual record cards, and placed a considerable burden upon the GRO. The LGB had also requested that the GRO undertake a supplementary inquiry to check the accuracy of the original survey.[107] This initiative may well have been linked to the riots of the unemployed in London in 1886, which

---

[102] PRO: RG 29/3, p. 144.
[103] Quoted in Eyler, *Victorian social medicine*, p. 192.
[104] GRO, *41st ARRG for 1878*; GRO, *42nd ARRG for 1879*.
[105] PRO: RG 29/3, p. 25.
[106] *Conditions of the working classes. Tabulation of the statements made by men living in certain selected districts of London in March 1887*. The topics covered in questions were – address; name; county of birth; marital condition; age; how long resident; if family resident; number of rooms occupied ; weekly rent; if physically equal to ordinary labour; disabled; in or out of work; weekly wages; time since last employment; cause of non-employment; means of subsistence when unemployed; length of time unemployed since 31 October; what members of family assist with income; weekly amount of such subsistence; name/relationship of informant.
[107] PRO: RG 29/3, pp. 25–28; MH 19/194, letter of 19 September 1887.

were to culminate in the battles of 'Bloody Sunday' in Whitehall in November 1887. These were to generate a veritable frenzy of private and public investigation into the lot of the East End poor in the following years.[108] Joseph Chamberlain, as President of the LGB, was so alarmed by the threat of the unemployed that in 1886 he issued a circular to the local authorities urging them to schedule necessary public works for periods of depression, and to co-operate with the Poor Law by providing paid, non-pauperising work for those who applied for poor relief due to temporary unemployment.[109] The GRO survey of 1887 plainly fitted into the Board's more pro-active stance.

The new relationship between the Board and the GRO's Statistical Department are clearly shown in an incident in the following year. On the 5 October 1888 Dr George Buchanan, the LGB's chief medical officer, asked the President of the LGB to request the GRO to send him copies of all death certificates where vaccination was the primary or secondary cause of death. An LGB official went to Somerset House and reported on 8 October that in the absence of Ogle he had seen Noel Humphreys, his deputy, who told him that Buchanan's request would involve an 'enormous expenditure of time' and doubted that 'the Registrar General would or could undertake the duty with his present staff.' Buchanan minuted Alfred Adrian, one of the Board's assistant secretaries, the next day to the effect that the President of the LGB, 'wants them, each and all, for the purpose of putting to the test the significance of anti-vaccination accusations'. He added that, 'these should, I think, be ordered off the Registrar General without his option'. Adrian, in turn, sent a minute to Henniker on 11 October telling him to 'request' Ogle and Humphreys to attend a meeting at the LGB, 'when in the event of the President being absent, the Board's medical officer, Dr Buchanan, will communicate the President's wishes on this matter.' They duly attended and 'agreed' to Buchanan's demands.[110] Similarly, the following year

---

[108] Jones, *Outcast London*, pp. 290–314.
[109] Thane, *The foundations of the Welfare State*, pp. 38–9. McBriar, *An Edwardian mixed doubles*, pp. 48–9.
[110] PRO: MH 19/195, correspondence of 5 October 1888 to 11 October 1888.

the LGB suggested that on registering a birth registrars should give the informant certain printed instructions with a view to checking on the prevalence of ophthalmia. The Registrar General indicated that he thought that this would be 'utterly futile for the purpose for which it is intended'. But since the president of the LGB wanted it done on a trial basis for a year, he undertook to circulate registrars asking them to distribute the memorandum, and the instructions were subsequently tested in a number of boroughs.[111]

It was this gradual realignment of forces within the LGB's sphere of influence which probably led to the increasingly 'bureaucratic and authoritarian ethos' of the Office in the late Victorian period, as noted by Szreter and others.

The limitations of Sir Brydges Henniker

Before leaving the subject of the inertia of the GRO in the period from 1880 to 1900, it is necessary to examine the role played in this by Sir Brydges Henniker. Henniker's lack of statistical training and his aristocratic background did not necessarily disqualify him for the post of Registrar General. Major George Graham had been from an equally landed background, and had seen military service, but had proved a successful and dynamic head of the Office. It should be noted, however, that Graham's experience as military secretary in colonial India probably placed greater administrative responsibilities on his shoulders than did Henniker's sojourn in the ranks of the Horse Guards. There is evidence, moreover, that Henniker was a weak man who allowed the GRO to drift for two decades.

Historians have tended to see Henniker's role mainly in terms of the failure of William Farr to become Registrar General in 1880 on the retirement of George Graham. The *Dictionary of National Biography*, for example, claims that, when Major Graham retired from the post in 1879, it had been generally expected that Farr would be appointed to succeed him. Farr himself is said to have desired to hold the post, if only for a short

[111] PRO: RG 29/19, pp. 2–3; RG 29/21, p. 52.

time. On the appointment being given to Sir Brydges Henniker, however, the *DNB* then claims that Farr resigned his post. It further claims that soon after his retirement paralysis of the brain set in, and that Farr died of bronchitis on 14 April 1883.[112] M. W. Flinn asserts that it was 'to the discredit of Victorian governments that he [Farr] was never appointed to the nominal headship of the department that owed everything to his genius, though the post became vacant more than once during his long career there'.[113]

However, this is a complete red herring. Farr's health had already been in decline before Graham's retirement. Indeed, the final letter the latter wrote to the Treasury, on 23 December 1879, was a request that Farr be allowed to retire on full pay, since 'his health has lately sadly failed in both body and mind, having obtained the age of 72 years'.[114] Putting himself forward as a possible Registrar General in these circumstances was a folly indicative of his failing powers, rather than a claim to his just deserts. In addition, the post of Registrar General was not a 'nominal headship' as Flinn claims but a demanding job that required considerable administrative expertise. As Eyler notes, Farr was not a man of business, and that even given robust health, he may not have been an ideal candidate for a major administrative post.[115]

However, there must be doubts as to whether George Graham's successor was up to the task either. Unlike Graham, Henniker often appeared unsure of his own powers, or even of the workings of the registration system, and frequently sought guidance from the LGB. In January 1882, for example, he sent a memo to the Board asking for legal advice on whether the 1874 Registration Act compelled doctors to provide certifications of cause of death on the printed forms introduced in 1845, or whether they could do so in another form. A minute on the memo from the LGB's legal advisor, D. P. Fry, seemed to indicate that the latter felt that Henniker was depending too heavily on the Board:

---

[112] *Dictionary of national biography* (London, 1889).
[113] Flinn, 'Introduction', p. 27.
[114] PRO: RG 29/2, p. 282.
[115] Eyler, *Victorian social medicine*, p. 191.

Before expressing any opinion, I think it is only right that the position of the matter should be clearly understood. The Registrar General can have no right to claim the Board's assistance in this way; and indeed he does not seem to claim it as a right, but merely to ask it as a matter of courtesy. Is it the wish of the Board that the advice should be given as a matter of courtesy?[116]

This might merely reflect the fact that Henniker had only been in post two years but such incidents continued over the following decade. In December 1886, for example, Henniker wrote to the LGB requesting legal advice on whether marriages could be celebrated in the presence of a deputy superintendent registrar. Alfred Adrian wrote to the Board's new legal advisor, J. F. Rotton, that, 'I should not have thought such a question as is here submitted to be sufficiently difficult or doubtful to require solution by the aid of the Law Officers.'[117] Again, in May 1891 the GRO wrote to the LGB asking it to adjudicate on what occupational terms should go under the heading 'living on their own means' in the census. In a minute on the file an LGB official noted, 'I should have thought that the Registrar General, who has the advantage of having all the documents before him, could better form an opinion upon this question than the Board.' He then went on to make some comments which were forwarded to the GRO.[118] Graham and Farr seldom showed such uncertainties, or similar deference to other departments.

Henniker appears to have been equally uncertain of himself in his dealings with his own staff, and with the Treasury. Whereas staff petitions for improved pay and conditions were, as already described, almost invariably forwarded by Graham to the Treasury with his support, Henniker's style was to forward such petitions whilst lamely declining to second them. Plainly, Henniker was trapped between the intransigence of the Treasury and burgeoning Civil Service trade unionism[119], but

---

[116] PRO: MH 19/194, letter of 16 January 1882.
[117] Ibid., letter of 14 December 1886.
[118] PRO: MH 19/195, letter of 11 May 1891.
[119] Wright, *Treasury control of the Civil Service*, pp. 227–31.

he appears to have rapidly lost the confidence of his own staff. When clerical discontent over pay and promotions boiled over in 1885, Henniker duly forwarded the clerks' petitions but indicated that they did not have his endorsement.[120] Instead he asked that three lower division clerks be given allowances of £20 to do superior work; for the number of 'duty pay places' to be increased as lower division clerks replaced assistant clerks; and for an external inquiry to consider the staffing structure of the Office.[121] The lower division clerks were so incensed by this that they wrote directly to the Treasury insisting that Henniker's suggested improvements to pay and conditions were 'totally inadequate to meet the just and reasonable expectations' of the GRO's staff.[122]

Nor does the Treasury seem to have been at all impressed by Henniker's handling of such matters. When in October 1885 Henniker forwarded a petition from his staff, the Treasury thanked him and then continued:

> My Lords desire me to state that, while they see no objection to your corresponding direct with this Board on matters connected with the estimates, or small details of your establishment, they do not think that they would be justified in considering any large scale measure of alteration in the scale of salaries in your department, unless it is submitted to them through the LGB, to which Board the Registrar General is by statute, in a measure, subordinate.[123]

The LGB subsequently declined to assume control over the Office's clerical establishment but the Treasury plainly felt that Henniker's lack of control needed to be pointed out.[124] It subsequently turned down Henniker's plea for a general enquiry into staffing structures.[125] The following year when

---

[120] PRO: RG 29/2, pp. 356–8, 363–6.
[121] Ibid., pp. 374–6.
[122] Ibid., p. 378.
[123] PRO: RG 29/6, p 282.
[124] Ibid., p. 284.
[125] PRO: RG 29/7, pp. 1–2.

Henniker suspended a lower division clerk for three months for falsifying returns of the number of certified copies he had completed, the Treasury over-ruled the Registrar General and insisted that the man be sacked.[126]

These events may explain Henniker's subsequent unwillingness to seek extra staff resources from the Treasury. The increasing workload of the GRO was met through ever-expanding overtime working, until by 1899 even the Treasury grew alarmed. In February of that year, Henniker indicated that approximately a third of the wages bill of the Office went on overtime payments, which led the Treasury to declare that, 'My Lords cannot but think that the figures given in your letter indicate the prevalence in your Department of a system of overtime employment during longer hours than is usual or in ordinary circumstances desirable in the public service.' They felt that this would impair efficiency, either through staff exhaustion, or via the temptation to clerks to perform their work during official hours in a slow or perfunctory manner in order to take advantage of extra overtime payments. The Treasury called upon Henniker to put forward plans for reducing the hours worked by his staff.[127] Given the Treasury's role in creating the problem in the first place, Henniker cannot be solely blamed for the situation. However, subsequent developments indicate that the Treasury might not have been unresponsive to earlier pleas for assistance. Henniker himself was unable to draw up a plan for reducing overtime working since he fell ill shortly after this exchange with the Treasury and retired by the end of the year.[128] In October 1899 the acting Registrar General, Noel Humphreys, asked for two additional second division clerks and one assistant clerk, a request to which the Treasury acceded the following month.[129] In the first years of the new century the staffing of the Office was to be expanded considerably.[130]

Henniker has, however, been credited with at least one success during his stewardship of the Office – that he was able

---

[126] PRO: RG 29/3, pp. 1–2; RG 29/7, pp. 1, 4.
[127] PRO: RG 29/3, pp. 219–20; RG 29/7, p. 134.
[128] PRO: RG 29/3, pp. 226, 241.
[129] Ibid., p. 236; RG 29/7, p. 143.
[130] See Chapter 5.

to maintain control of the decennial enumeration during the deliberations of the 1890 Treasury Committee on the Census. The latter arose out of a deputation of social scientists, led by Charles Booth and the economist Alfred Marshall, to the Chancellor of the Exchequer and the President of the LGB in 1888. This was itself an indication of the increasing institutional subordination of the GRO. The formal demands of the deputation included:

> a quinquennial census;
>
> that the census should be the 'care of a special department' which should be in continuous existence;
>
> that a question on employment status should be added to the census schedule; and
>
> that there should be revisions to the occupational classification.[131]

Simon Szreter has portrayed this deputation and the resulting Committee as an attempt by social scientists and economists to wrest control of the occupational census from the GRO. Booth and Marshall were attempting to shift its underlying organisational principles away from Farr's medical model of the effects of working with materials on health, to those more suitable for economic and social analysis.[132] Szreter sees the GRO as opposing this change on the grounds of the work involved, the need to maintain the consistency of its occupational classifications systems over time, and a desire to ensure its ability to match the occupational information in the census with that derived separately from the death registration system.[133] However, it should be noted that the GRO had indeed made rather extensive changes to the occupational classification system used in the census in 1881, and that the resulting lack of compatibility between this census and that of earlier decades was one of

---

[131] Enclosure to PRO: T 1/8487B/9295.
[132] Szreter, *Fertility, class and gender in Britain*, pp. 114–20.
[133] Ibid., pp. 117–8.

Charles Booth's main criticisms of the Office.[134] The appointment of the Treasury Committee was plainly a very important challenge to the autonomy of the GRO, especially given the presence on the Committee of Sir Reginald Welby, the Treasury's permanent secretary. Szreter sees the Committee's deliberations as having gone well for the GRO, in that its recommendations mirrored those put forward by the GRO, bar the Committee's support for the inclusion of an employment status column that he sees as a 'very partial victory for Marshall and Booth'. In addition, Szreter argues, Welby seems to have been satisfied that there was no serious extravagance to be found in the GRO and that its management of the census should continue.[135]

There are, however, some problems with this argument, and it is difficult to use the events surrounding the Committee to put Henniker's husbanding of the census and GRO in a good light. The fact that social scientists were so unhappy with the census, and that the Treasury should have taken their complaints seriously enough to set up a committee of enquiry does not indicate great faith in Henniker. The actual recommendations of the Committee were not dissimilar to the demands of the social scientists' original deputation, being

> a quinquennial census;
>
> that a 'small permanent census branch of the department of the Registrar General' should be established;
>
> a permanent Census Act, rather than a temporary one every ten years;
>
> the introduction of a question on employment status;
>
> the omission of rank from the schedule heading 'rank, profession or occupation';
>
> and, the introduction of a question on the number of rooms inhabited if less than five.[136]

---

[134] Higgs, *A clearer senses of the census*, pp. 158–9.
[135] Szreter, *Fertility, class and gender in Britain*, p. 115
[136] *Report of the Treasury Committee on the Census*, p. xii.

The GRO itself supported the first three recommendations, and they indeed hung together since if there was a census every five years there would be work for a permanent census establishment, and a permanent Census Act would be necessary for its funding.[137] A quinquennial census was sought by both the GRO and MOHs because it would allow the more accurate calculation of local death rates. The base population used by the GRO for calculating a local mortality rate in 1890 was estimated by applying the logarithm of the local population growth found between the censuses of 1871 and 1881 to the enumerated population in 1881. Plainly, if the rate of population growth after 1881 was greater or less than that between 1871 and 1881, the estimated population in 1890 would be either too large or too small, thus affecting the accuracy of a calculated death rate per 1000 population.[138] But Welby argued against these proposals in a minority report on the grounds of cost, and they were not implemented.[139] Cannily, Welby suggested instead a system whereby local councils could pay for a quinquennial census if they wanted one. The GRO thus had the worst of all possible outcomes – it had to respond to the distraction of demands for local enumerations, without having the extra permanent resources to undertake them.[140]

Similarly, the GRO opposed any extensions to the census schedule in the late nineteenth century but it was just these aspects of the Committee's recommendations that were adopted by the government and introduced in the 1891 census, at the insistence of the LGB and despite the GRO's continuing opposition. Ogle tried to claim that the Census Act did not give the GRO authority to collect the data – an interpretation which was resisted by the LGB.[141] Although the GRO declined to analyse the data on employment status collected in 1891, since

[137] Ibid., Appendix 6, p. 121. The GRO had asked the Treasury for funding for a permanent census office in 1887 and had been turned down: PRO: RG 29/3, p. 46.
[138] Mooney, 'Professionalization in public health', pp. 57–8.
[139] *Report of the Treasury Committee on the Census*, pp. xiii–xiv.
[140] For the surveys carried out on this basis, and the difficulties they created for the GRO, see PRO: RG 19/42–44B.
[141] PRO: MH 19/195, letters of 30 August, 8 September, and 15 September 1890; Schürer, 'The 1891 census and local population studies', pp. 16–29.

it claimed that the returns were unreliable, the question reappeared, in a modified form, in 1901.[142] As Szreter notes the GRO henceforth included representatives from other government departments in internal committees planning the census.[143] It was the Board of Trade, for example, that suggested the splitting of the occupation column in two, for occupation and the trade or industry in which the occupation was performed, which was a feature of the 1911 census.[144] Even Marshall's suggestion for the establishment of a unitary census-taking department was not a serious attack on the Office, since he seems to have been quite happy for that department to be in the GRO.[145] On the whole, the outcome of the 1890 Committee was not a success for the GRO – the establishment of a permanent structure for taking the census was put back a generation[146], and the Office lost its exclusive control over the form of the decennial enumeration.

Conclusion

There were plainly problems in the GRO in the final decades of the nineteenth century. Of the reasons for this inertia that scholars have put forward, intellectual factors are the most interesting if the most difficult to prove. More mundane institutional issues of staffing, leadership and Whitehall control may, however, have been of equal importance. The changing relationship between the GRO and its parent department, especially its increasing involvement in the activities of the LGB, was undoubtedly a factor in the gradual re-orientation of the Office's interests. Rather than focusing on forms of social reciprocity based on the family and locality, the GRO increasingly

---

[142] Higgs, *A clearer sense of the census*, pp. 108–9.
[143] Szreter, *Fertility, class and gender in Britain*, p. 119–20.
[144] RG 19/48B Departmental committee on the census of 1911, p. 103.
[145] *Report of the Treasury Committee on the Census: Evidence*, Q. 1517.
[146] The GRO continued to press for a quinquennial census and a permanent census office: PRO: RG 19/42 Census of London 1896; proposed censuses in 1906 and 1916: correspondence, letter of 24 October from the chief clerk to the LGB; PRO: RG 19/48A, 1911 Census Bill, departmental memo to the president of the LGB, 11 February 1910.

found itself drawn into underpinning central policy formation and implementation within Whitehall.

This structural context needs to be integrated, however, with the very personal limitations of Brydges Henniker. If the GRO lost its way in this period; if it was unable to impress on the Treasury the urgency of its staffing problems; if it was unable, or unwilling, to assert its freedom of action vis-a-vis its parent department, then some responsibility must attach to the personality and outlook of the Registrar General in the last two decades of the nineteenth century. These limitations were to be thrown into relief by the striking reforms within the Office after Henniker's retirement.

5

1900–1914: eugenics and the GRO's Indian summer

The evidence of statistical revival

Whatever the nature and causes of the GRO's apparent inertia in the late Victorian period, there is no doubting the appearance of a new vitality and expansiveness in the Office's statistical activities in the early years of the twentieth century. As can be seen from Figure 3:2, the size of the GRO's *Annual reports* began to grow rapidly from the turn of the century. The *Annual report* for 1901, published in 1903, reinstituted the distinction between the Registrar General's report and the Superintendent of Statistic's 'Letter'. Sir William Dunbar, the Registrar General from 1902 to 1909, justified this change in presentational and professional terms, stating that he had:

> adopted this course as it occurs to me that the medical practitioners of this country, on whose generous co-operation the accurate compilation of vital statistics so largely depends, will, in this way, more readily appreciate the fact that the particulars they contribute concerning the causes of death are analysed, and the results authenticated, by a member of their own profession.[1]

In his Letter in this volume, John Tatham noted the more general introduction of death rates standardised by age in the attached tables, which had been made necessary because of changes in the age structure of the population consequent upon the decline in the birth-rate in the late nineteenth century. This

---

[1] GRO, *64th ARRG for 1901*, p. xxi.

made comparisons of crude death-rates with those of earlier periods misleading.[2] The following year he was pointing to improvements in the GRO's statistical output in terms of supplementary tables on infected organs and parts of the body, and on deaths in childbirth.[3] In the *Annual report* for 1905 the Office began publishing tables on a continuous basis showing mortality from the principal causes of death in each of the first four weeks after birth, and in each month in the first year of infant life.[4] That for 1906 saw the re-introduction of graphs to illustrate statistical themes.[5] In 1911 the Superintendent's Letter became the body of the *Annual report* for 1909 under the title 'Review of Vital Statistics', with the Registrar General's report forming an introduction of a mere four pages. According to Sir Bernard Mallet, the Registrar General from 1909 to 1920, this was to enable the statistics to be treated as a whole, without the overlapping and repetition common under the former system of dividing the *Annual reports* between two authors.[6]

T. H. C. Stevenson, the GRO's chief medical statistician from 1909 to 1931, also created the first coherent model of socio-economic groupings, based on the placing of the occupations of household heads into an hierachical classification system, for the 1911 census.[7] This analytical model was first applied by Stevenson to marital fertility in the 1911 enumeration[8], and then to mortality[9], before going on to become a standard tool

---

[2] Ibid., p. xvi.
[3] GRO, *65th ARRG for 1902*, p. xxxiv.
[4] GRO, *68th ARRG for 1905*, pp. cxviii–cxxxiii. From 1889 onwards the GRO had published tables showing infant mortality in the first and second three month periods of life, and in the second six month period: GRO, *51st ARRG for 1888*, p. ix. For other innovation in this period with respect to infant and child mortality, see: Szreter, *Fertility, class and gender in Britain*, pp. 247–8. See also, Armstrong, 'The invention of infant mortality'.
[5] GRO, *69th ARRG for 1906*, p. lxiv.
[6] GRO, *72nd ARRG for 1909*, p. v.
[7] Szreter, *Fertility, class and gender in Britain*, pp. 238–82.
[8] GRO, *75th ARRG for 1912*, pp. viii, xxii–xxxii; *Census 1911, Vol. XIII, Fertility of marriage report*, Part 1; *Census 1911, Vol. XIII, Fertility of marriage report*, Part 2; Stevenson, 'The fertility of various social classes'.
[9] GRO, *Supplement to the 75th annual report of the Registrar General: Part IV*, pp. i–xii, 3–6; *The Registrar-general's Decennial Supplement, England and Wales 1921. Part II*, pp. v–lii; Stevenson, 'The social distribution of mortality from different causes in England and Wales'; Stevenson, 'The vital statistics of wealth and poverty'.

Eugenics and the GRO's Indian summer 131

of social science research.[10] The introduction of questions on completed married fertility into the 1911 census was itself a major departure since for the first time householders were asked to provide information other than on the characteristics of household members present on census night. The 1911 questions on fertility asked information for each married woman on the total number of children born alive to the present marriage, the number still living, and the number dead.[11] In addition, the *Annual reports* published between 1911 and 1913 saw a considerable number of innovations – the publication of mortality data by sanitary rather than registration districts; the consistent return of deaths in hospitals to place of usual residence for the purposes of tabulation; the abandonment of the GRO's internal nosologies and the introduction of the abstraction of deaths according to the International List of Causes of Death[12]; greater detail respecting the cause of death in relation to age and sex for aggregates of large towns, smaller towns, and rural areas, which had hitherto only been given for the country as a whole and for London; new tables giving a double classification of mortality by primary and secondary causes; and the introduction of data on mortality according to place of death – private house, hospital, workhouse, and so on.[13]

The GRO's Statistical Department appears, therefore, to have been undergoing a veritable renaissance in these years. This chapter will consider the direct impact of eugenics on this process, whilst the next will consider some of the indirect, and perhaps ultimately more important, effects of this new discipline on the work of the GRO.

---

[10] Szreter, 'The genesis of the registrar-general's social classification of occupations'.
[11] For the form of the 1911 census schedule, see: Szreter, *Fertility, class and gender in Britain*, pp. 604–5.
[12] For the development of the International List as an internationally accepted standard nosology, see: Moriyama, Loy, and Robb-Smith, *History of nomenclature of diseases*.
[13] GRO, *72nd ARRG for 1909*, pp. vi–viii; GRO, *73rd ARRG for 1910*, pp. vii–viii; GRO, *74th ARRG for 1911*, p. x.

Mortality statistics and eugenics

In line with his thesis regarding the intellectual causes of the GRO's inertia in the late Victorian period, Simon Szreter sees the struggle between eugenicists and public health environmentalists as the key to understanding this new-found dynamism:

> In an entirely unintended way the hereditarian eugenicists restored the scientific and political significance of the GRO because their methodology once more returned issues surrounding the interpretation of demographic and epidemiological statistics to centre stage . . . the GRO's traditional expertise in statistical inference and as official custodian of the nation's demographic record once more returned the Office into the political limelight, as it provided the public health movement with its most effective intellectual resources for a refutation of the eugenicists' views – in their own 'scientific' and statistical terms.[14]

He points to the close working relationship between Sir Arthur Newsholme as chief medical officer of the LGB, and Stevenson at the GRO. Stevenson had actually been Newsholme's deputy as MOH in Brighton[15], and provided the latter with data for his public sparring with eugenicists such as Karl Pearson. Newsholme drew extensively on GRO data in the years 1910 to 1914 to show that infant mortality was declining, and that high infant and high child mortality were associated. These arguments undermined eugenic assertions that high infant mortality reflected the birth of 'weak' children, the off-spring of degenerate stock preserved from 'natural' evolutionary elimination by misguided environmental improvements. If mortality was still high amongst those working-class children who had survived the ravages of infantile diseases, this pointed to a hostile environment rather than to the elimination of the 'unfit.'[16]

---

[14] Szreter, 'The GRO and the public health movement in Britain', pp. 458–9.
[15] This was a fact which Stevenson stressed in his letter of application for the post of Superintendent of Statistics at the GRO: PRO: T 1/11014/8958, letter of 25 March 1909.
[16] Eyler, *Sir Arthur Newsholme and state medicine*, pp. 297–305.

An apparent problem with this explanation of the GRO's statistical renaissance is that Szreter, as has already been noted, uses the eugenicist onslaught to explain both the *contraction* of the GRO's activities in the late Victorian period and their *expansion* in the early twentieth century. Why should eugenics have had such a different impact on the activities of the GRO in the two periods? His argument appears to be that the difference between the two episodes was that in the earlier period environmentalists lacked a model of social class which would enable them to show how their emphasis on administrative action to remove physical nuisances, or to educate the population, could bring lasting and sustainable improvements. Szreter argues that 'by the opening years of the new century the country's leading public health officials and salaried educationists were gradually piecing together a perception of the detailed operation of certain impersonal and economic forces acting within and upon the country's poorer communities: self-reinforcing circles of impoverishment and disadvantage upon the poor.'[17]

Szreter points, for example, to the evidence of Dr Alfred Eicholz before the Interdepartmental Committee on Physical Deterioration. Eicholz, an inspector in the Medical Department of the Board of Education, 'saw the structured labour market as an encompassing, repetitive social process which systematically inflicted differential environments and life experiences on various parts of the population.' He held that physical deterioration reflected poor diet, clothing and housing, rather than bad genes.[18] As Szreter shows convincingly, Stevenson contributed to this developing environmentalist counter-attack by constructing his classification of socio-economic groupings, and using it to argue that the fertility transition to smaller families represented the spread of knowledge of contraceptive techniques from the more 'intelligent' classes 'downwards'.[19] Environmentalist improvements would not lead, therefore, to racial degeneration, since the diffusion of birth-control methods would equalise class fertility rates.

[17] Szreter, *Fertility, class and gender in Britain*, p. 234.
[18] Ibid., pp. 212–15.
[19] Ibid., pp. 254–71.

Szreter argues, moreover, that the creation of this model of social class reflected a long-standing interest on the part of the officers of the GRO in the subject. In order to substantiate this argument he points to a paper given to the Royal Statistical Society in 1887 by Noel Humphreys, one of the GRO's principal officers, on 'Class mortality statistics'.[20] This surveyed extant attempts to analyse differentials in class mortality rates, especially the experimental social classification of occupations developed by the Irish Registrar General, Dr T. W. Grimshaw, for Dublin, and called for similar work on England.[21] Szreter then proposes that the GRO's contemporary survey of the working classes in the East End of London was another example of a nascent interest in social class.[22] He notes, however, that after this flurry of activity the GRO then 'strangely lost interest in social class analysis', concentrating instead on improving its existing spatial and temporal analysis of infant and child mortality.[23]

Mortality statistics and the population question

Szreter's argument is extremely well argued and persuasive, and is plainly very important in explaining crucial aspects of the development of the GRO's activities in the period just before the First World War. There is a sense, however, in which placing the GRO's revival in the context of a Manichean struggle between eugenicists and public health environmentalists obscures as much as it reveals. The renewed national interest in population and questions of infant and child mortality in the early twentieth century, in which the GRO shared, was not simply due to the activities of the eugenicists. The declining birth rate implied a problem with respect to the maintenance of the *quantity* of the British population in an age of imperial rivalry, as much as with its *quality*. The registered birth rate had fallen from 35.33 per thousand in the decade 1871–80 to 26.59

---

[20] Ibid., pp. 78–80.
[21] Humphreys, 'Class mortality statistics'.
[22] Szreter, *Fertility, class and gender in Britain*, pp. 81–2.
[23] Ibid., p. 85.

Eugenics and the GRO's Indian summer 135

in the period 1901–11[24], whilst the infant mortality rate had failed to improve in the late Victorian period in line with the experience of other age cohorts. As John Tatham noted in the *Decennial supplement* for 1891 to 1900, published in 1907:

> it appears that, although in the course of the last four decennia the death rate of all ages has fallen by 15 per cent., and the death-rate at ages one to five by not less than 33 per cent., nevertheless at ages under one year the death-rate in 1891–1900 has shown no reduction from the high rate recorded in 1861–70. The infant portion of the community has not shared in the common benefit.[25]

Edwin Cannan had argued in the *Economic Journal* in 1895 that the declining replacement rate would lead to a stationary population, whilst those of Germany and the USA were expanding rapidly.[26] Such an eventuality would have dire consequences for British society, the economy, and the Empire.

This led to the development of a general child-welfare movement that sought to prevent the loss of infant life. As historians have argued, predictions with respect to the declining size of the population encouraged a public debate over infant and maternal welfare, especially after the debacle of the Second Boer War.[27] This led the President of the LGB, John Burns, to hold National Conferences on Infant Mortality in 1906 and 1908. The opening addresses that Burns gave to both conferences, which were attended by large contingents of MOHs, were distinctly bereft of allusions to eugenic ideas. He explained the 'grisly fact of infants dying before their time' in terms of the ignorance of mothers, drunkenness, and of women working

---

[24] Mitchison, *British population change*, p. 24.
[25] GRO, *Supplement to the 65th annual report of the Registrar General*, p. cv.
[26] Nissel, *People count*, pp. 129–30. The infant mortality rate in fact improved in the Edwardian period, declining from 153 per thousand in 1891 to 1900 to 128 per thousand in 1901–10: GRO, *Supplement to the 75th annual report of the Registrar General*, p. xxxiii.
[27] Lewis, *The politics of motherhood*; Dwork, *War is good for babies* ; Brand, *Doctors and the State*, pp. 177–83; Thane, *The foundations of the Welfare State*, pp. 60–1 .

outside the home.[28] After the First World War, under the 1918 Maternal and Child Welfare Act, each local authority was to establish a maternal and child welfare committee, and it enabled, but did not compel, them to set up a maternal and child welfare service.[29] The belief in 'low maternal efficiency' – the presumed fecklessness and ignorance of cleanliness amongst working-class women – led to attempts to educate women in motherhood.[30] Such measures were linked to a pro-natalism that even swept the eugenics movement along in support of general infant welfare programmes to replace men lost during the War.[31] Although the rhetoric of child welfare, with it emphasis on improving the quality of the race, can be seen as eugenic, its general aim of preserving the lives of infants could not be squared with the policy of the survival of the fittest advocated by extreme eugenicists.

The crucial issue here is the extent to which the GRO's heightened interest in infant and child welfare immediately prior to the Great War was spurred by the concerns underlying the general child-welfare movement, or more specifically by struggles with the eugenicists. Certainly, the GRO now began to take infant mortality very seriously. The Office had addressed the issue sporadically in the nineteenth century[32] but the subject took on a new centrality in the early years of the twentieth century. As well as the new tables already noted, a distinct section on mortality amongst infants and young children was introduced as a standard component of both the *Annual reports* and *Decennial supplements* in these years.[33] In the 1920s such mortality was usually the first subject to be discussed in the annual *Statistical review.* The GRO's officers appear initially, however, to have seen the introduction of the more detailed

---

[28] Burns, 'Presidential address delivered to the First National Conference on Infantile Mortality'; National Conference on Infant Mortality, *Report of the proceedings of the National Conference on Infantile Mortality,* pp. 11–28.
[29] Lewis, *The politics of motherhood,* p. 34.
[30] Ibid., pp. 61–113.
[31] Soloway, *Demography and degeneration,* pp. 136–62, 172–5.
[32] See, for example, GRO, *38th ARRG for 1875,* pp. xl–lii; GRO, *40th ARRG for 1877,* pp. xxvii–xxx; GRO, *54th ARRG for 1891,* pp. x–xxiii.
[33] GRO, *67th ARRG for 1904,* pp. xciii–xcviii; GRO, *Supplement to the 65th annual report of the Registrar General,* pp. cv–cxvi.

analysis of infant mortality in terms of an administrative response to the 1906 National Conference on Infant Mortality. According to John Tatham writing in the *Annual report* for 1905, published in 1907, although the fact that infant mortality had not declined over the previous half century:

> has been kept steadily before the public in the Registrar General's successive reports, it is only in comparatively recent times that public interest has been thoroughly awakened. Last autumn, however, the President of the LGB called a conference of authorities to consider this question, and the medical officer of the Board has since taken steps to procure periodically from medical officers of health throughout England and Wales special returns of mortality in the infantile portion of the community.[34]

As Tatham noted in his Letter the following year, the form of the new GRO tables conformed to that in the MOHs' reports required by the LGB, 'so that MOHs have at hand reliable standards with which to compare the infant mortality in their own districts.'[35] Similarly, Dunbar emphasised the role of the 1906 National Conference on Infant Mortality in awakening public interest in the subject of the 'wastage of infant life'.[36] Dunbar's general view was that since, 'little can probably be done by legislation to arrest the increasing decline in the birth-rate, it is no doubt of paramount importance that effective measures should be devised to lessen the enormous death-toll caused by "Infant Mortality".'[37] In correspondence with the Treasury, Dunbar again justified his use of staff on overtime on the new infant mortality tables in terms of the need to provide MOHs with data to compare to the annual reports on infant mortality which the LGB had recently asked them to submit.[38]

It was only when Sir Bernard Mallet took over as Registrar

[34] GRO, *68th ARRG for 1905*, p. cxviii.
[35] GRO, *69th ARRG for 1906*, p. lxix.
[36] GRO, *69th ARRG for 1906*, p. lxv.
[37] GRO, *67th ARRG for 1904*, p. xlviii.
[38] PRO: RG 29/3, p. 367.

General in 1909 that the problem of infant mortality began to be seen more specifically in terms of the quality of the population. Thus, when discussing the introduction of tables in the *Annual report* for 1911 which cross-tabulated infant mortality by parental occupations, Mallet noted that, 'the extent of their bearing upon the prevention of infant mortality may prove to be a controversial matter, depending upon the relative importance attributable to inheritance and environment as factors causing mortality, but at least it must be regarded as very considerable.'[39] The following year he used his report to draw attention to Stevenson's presentation in his appended Letter of data on births by occupational and social groups, noting that, 'fertility varies greatly with social status, being with few exceptions lowest for professional and other middle-class occupations and highest as a rule for those representing unskilled labour'.[40] The increased salience given to such population issues in the Edwardian period appears to have sprung, therefore, from a mixture of concerns, of which the struggle over eugenics was not necessarily the most important.

The eugenicist-environmentalist polarity is also one that can be overdrawn. Greta Jones has argued, for example, that nineteenth-century environmentalism and eugenics came together in the early twentieth century in the concept of 'social hygiene'. This was a marriage between the social darwinism that stressed heredity, fitness and racial progress, and the practical techniques of social management that evolved from the nineteenth-century tradition of public health and charity work. Social hygiene was influential precisely because it was easily integrated into the discourse of nineteenth-century health reform. The public health reformers of the nineteenth century urged the poor to be thrifty, far-seeing, hard working as well as physically and mentally healthy. At the same time they had discovered the 'residuum' – those amongst the working classes whom no admonition, advice, discipline or instruction seemed able to save from unemployment, immorality and ill health. These concepts

---

[39] GRO, *74th ARRG for 1911*, p. xi.
[40] GRO, *75th ARRG for 1912*, pp. viii, xxii–xxxii.

Eugenics and the GRO's Indian summer

easily dove-tailed with that type of social darwinism which divided the nation into the fit and the unfit and which, although it held that some of the unfit could be made fitter, attributed the existence of the residuum largely to heredity.[41] This was certainly how Charles Booth distinguished between his residual 'classes A and B' and other components of the working classes in his *Life and labour of the people in London*.[42] Eugenics and public health environmentalism were probably not as coherent and exclusive in ideological terms as they have sometimes been portrayed.[43]

The relationship between the GRO and the eugenicists was also not as adversarial as one might suppose. Certainly the eugenicists took the results of the 1911 fertility census, which did indeed reveal class differentials, as confirmation of their own arguments.[44] Similarly, in 1923 Stevenson could happily discuss his analysis of the relationship between social class and mortality, as recently published in the GRO's *Decennial supplement*, before Pearson's Society of Biometricians.[45] Five years later Sir Bernard Mallet, who resigned as Registrar General in 1920, was to become the president of the Eugenics Society. Szreter argues that Mallet was not interested in demographic matters whilst at the GRO, and only became an advocate of eugenics after leaving the Office. He bases this on the fact that Mallet's major *published* work whilst Registrar General related to British budgetry and fiscal policy.[46] As noted above, however, Mallet did indeed appreciate the eugenic implications of the GRO's work on infant mortality and class fertility in the very years when the debates between Newsholme and Pearson were at their fiercest. Mallet's analysis of British budgets in the period 1887 to 1913 also had a direct eugenic import, because he was

---

[41] Jones, *Social hygiene in twentieth century Britain*, p. 160. For the continuity between eugenics and Victorian social policy, see: Mazumdar, *Eugenics, human genetics and human failings*, pp. 37–8.
[42] Fried and Elman (eds), *Charles Booth's London*, pp. 9–19.
[43] For similar arguments, see: Freeden, 'Eugenics and progressive thought', pp. 645–54; Thomson, 'Sterilization, segregation and community care', pp. 476, 492.
[44] Soloway, *Demography and degeneration*, pp. 228–9.
[45] GRO, *Supplement to the 75th annual report of the Registrar General: Part IV*; Stevenson, 'The social distribution of mortality'.
[46] Szreter, *Fertility, class and gender in Britain*, p. 249, n. 49; Mallet, *British budgets*.

attempting to understand how changes in taxation and government expenditure impacted on the social income of the differing classes.

Moreover, as will be discussed in Chapter 6, Stevenson's introduction of questions on marital fertility into the 1911 census had profound implications for the staffing and work of the Census Office, and for the mechanics of data handling. It was Mallet, rather than Stevenson, who had to work hard to persuade the Treasury to accept these innovations, which implies that he appreciated the importance of the fertility survey. Again, one must not assume because administrators do not publish on a particular subject that it is not of professional concern to them. In addition, if Mallet was not interested in eugenics whilst at the GRO, then one must also assume that his literary interest in the subject developed very suddenly after his resignation as Registrar General at the end of 1920 , since he published an essay entitled 'Is England in danger of racial decline?' as early as February 1922.[47] As will be discussed in Chapter 7, Stevenson's successor as medical statistician in the GRO was none other than the medical officer of Karl Pearson's Eugenics Laboratory at UCL.

It should also be noted that eugenics did not fail to make significant headway in Britain in the period before the First World War simply because of the elaboration of an alternative model of the relationship between social class, fertility and mortality. Rather, eugenics ran counter to the entrenched liberal concepts of the nature and responsibilities of citizens. At the heart of much of the adverse reaction to eugenics was the belief that it removed moral responsibility from the individual for his or her own predicament. Individuals could not be blamed for failing to maintain themselves or their families, or for being unequal to the obligations placed upon them by society, since this merely reflected their genetic inheritance. Opposition to the idea that individuals were genetically incapable of improving themselves united Sir Arthur Newsholme, MOHs, and

---

[47] Mallet, 'Is England in danger of racial decline?'. In the same year Mallet also published, 'Registration in relation to eugenics'.

bodies such as the Charity Organisation Society, against eugenics.[48] Even that apparently most eugenic of measures, the detention and institutionalisation of the mentally deficient under the 1913 Mental Deficiency Act, was introduced in terms of protecting the individual mental defective against society, and was thus shorn of eugenic implications. To conceive of people as the bearers of genes rather than as responsible members of a social system with rights and duties was to remove their very humanity — hence the constant charge that eugenics reduced society to the level of the cattle market.[49] The eugenicists' emphasis on disadvantaging the poor also ran counter to the strategy of re-attaching marginal or disaffected groups to the social order which underlay so much British social policy in the Victorian and Edwardian periods. Eugenics was also not practical party politics in an age when the working classes were becoming the majority of the electorate.[50]

The GRO and socio-economic class

Stevenson's introduction of questions on marital fertility into the 1911 census, and his creation of the model of socio-economic groupings, was certainly in order to test the validity of eugenic arguments. As he explained to an internal GRO committee, 'The upper and middle classes . . . were constantly being accused of not reproducing themselves, and he thought it desirable that statistics should verify or deny this accusation.'[51] However, one may question the centrality of socio-economic class to the broader interests of the GRO. Despite attempts to provide a lineage for Stevenson's innovation in the work of Farr, Ogle and the Victorian GRO, the evidence to support this view is problematic. Humphreys' 1887 essay on social class and

[48] Searle, *Eugenics and politics in Britain*, p. 64; Porter, 'Enemies of the race'; Eyler, 'The sick poor and the state', pp 202–6.
[49] Freeden, *The New Liberalism,* pp. 190–4; Larson, 'The rhetoric of eugenics; Thomson, 'Sterilization, segregation and community care'.
[50] MacKenzie, *Statistics in Britain,* p.50; Davidson, *Whitehall and the labour problem,* pp. 247–8.
[51] PRO: RG 19/48B Departmental Committee on the Census of 1911 (Oct. 1909–Apr.1911), p. 62. He repeated this in a paper to the Royal Statistical Society in 1910: Stevenson, 'Suggested lines of advance in English vital statistics', p. 705.

mortality was indeed a path-breaking work, although even he saw the problem in administrative rather than socio-economic terms. As he put it:

> poverty and hard work do not necessarily kill; but only when poverty and hardships are accompanied by dirty houses, impure air, and above all by intemperance, which so frequently prevails among badly housed populations. . . . In the face of such facts as these, it is impossible to doubt that much of the large excess of mortality among the working classes is distinctly within the control of effective sanitation.[52]

But Humphreys went on in his paper to indicate how Farr and Ogle's official work on life-tables and occupational mortality was unsuitable for an analysis of class mortality, especially given the concentration in the latter on adult males rather than infants.[53]

Farr was more interested in the effects of population density on mortality than in social class *per se,* whilst Ogle's main concern was with occupational health. This can be seen from the prominence Ogle gave to it in the *Decennial supplement* he published in 1885, and in his reworking of the occupational classification system used in the 1881 census.[54] Indeed, he appears to have seen occupational morbidity and mortality as more important than local public health issues. As he told the Seventh International Congress of Hygiene and Demography in 1891:

> Favourable and unfavourable climatic and geographical conditions, activity or inactivity on the part of the sanitary authorities to provide wholesome water, to remove filth, and to prevent overcrowding, contribute doubtlessly in no small measure to reduce a death-rate in one town, or to raise it in another; but all differences so produced, are insignificant,

---

[52] Humphreys, 'Class mortality statistics', p. 279.
[53] Ibid., pp. 256, 265.
[54] GRO, *Supplement to the 45th annual report of the Registrar General*, pp. xxi–lxiv; *Census of England and Wales 1881. Vol. IV. General report*, p. 26; Higgs, *A clearer sense of the census*, pp. 154–67.

when compared to the differences shown by death rates in different occupations.⁵⁵

Ogle saw various factors as contributing to occupational mortality differentials – cramped or uncramped working conditions, relative exposure to poisons and dust, workload, the degree of ventilation, and differential liability to fatal accidents.⁵⁶ Under the impetus of Ogle's work, John Tatham also produced a whole volume on occupational mortality as part of the *Decennial supplement* for 1881 to 1890.⁵⁷

Humphreys' paper was, moreover, poorly received by the members of the Royal Statistical Society. Most speakers in the discussion afterwards criticised Grimshaw's classification, and argued in true English fashion that Dublin was so peculiar a place that one could not make general inferences from it. F. G. P. Neison doubted the existing tables showed any difference between the classes, and stressed that the GRO should concentrate on differences of mortality between localities and differing occupations. Dr James Edmunds, MOH for St James, put down high infant mortality in Dublin to damp housing on clay, and the consumption of alcohol and tobacco amongst the poor.⁵⁸ No other member of the staff of the GRO appears to have been present at the meeting to speak in Humphreys' defence. As already noted, Szreter's other item of evidence for the GRO's interest in social class in this period, Ogle's 1887 survey of working-class conditions in the East End, was not an internally generated 'research initiative' but an administrative task laid on the GRO by the LGB. Szreter sees this survey as an internal GRO initiative *preceding* the interest of the rest of government in the subject of unemployment and social class because he incorrectly dates the survey to 1884, and appears unaware of the LGB's role in its initiation.⁵⁹

⁵⁵ International Congress of Hygiene and Demography, *Transactions of the Seventh International Congress of Hygiene and Demography*, Vol. X, Div. II , p. 12.
⁵⁶ Ibid., pp. 15–22.
⁵⁷ GRO, *Supplement to the 55th annual report of the Registrar General*.
⁵⁸ Humphreys, 'Class mortality statistics', pp. 285–92.
⁵⁹ He appears not to have used the published results of the survey in the Parliamentary Papers, relying on a text delivered by Ogle to the third meeting of the International Statistical Institute in Vienna in 1891; Szreter, *Fertility, class and gender in Britain*, pp. 81–2.

There was nothing strange, therefore, in the GRO failing to follow up an official programme of socio-economic research that it never really had, and in the 1890's the Office continued its traditional role of analysing natality and mortality in terms of locality, settlement type, population density and occupation. In this the GRO was falling behind countries such as Denmark, where Theodor Sorensen had been publishing data on the effects of social class on infant mortality since 1883.[60] Stevenson's work in this field should be credited, therefore, as an innovative break with the past rather than as the culmination of the GRO's research efforts.[61] In this respect it is also interesting that Noel Humphreys opposed the introduction of Stevenson's fertility survey into the 1911 census on the grounds that it would overburden the census schedule.[62] Given that this new departure in the work of the GRO coincided with the career within the Office of a Registrar General with incipient eugenic interests, one might equally see the concept of socio-economic groupings as having roots in this pseudo-science.

The elaboration of Stevenson's social class model cannot be used to explain much of the expansiveness of the GRO in the period between 1900 and the First World War. It hardly provided public health environmentalists in these years with a coherent answer to the eugenicists' claims regarding the adverse effects of differential class fertility rates on national health. As Szreter has shown, Stevenson was initially unhappy with his creation, and very nearly abandoned it in favour of social criteria such as the numbers of rooms occupied, or domestic servants employed, per family in his analysis of the 1911 fertility data.[63] As already noted, the results of the 1911 fertility survey did indeed reveal class differentials in fertility that appeared to support the eugenicist case.[64] As Stevenson confessed in 1914, when discussing the preliminary results of applying his socio-

[60] Lokke, 'No difference without a cause', p. 90.
[61] That Stevenson initiated the GRO's interest in the effect of social class on mortality was also the view of Major Greenwood: Greenwood, 'The occupational and economic factors of mortality', p. 864.
[62] Stevenson, 'Suggested lines of advance in English vital statistics', p. 704.
[63] Szreter, *Fertility, class and gender in Britain*, p. 258.
[64] Soloway, *Demography and degeneration*, p. 17.

Eugenics and the GRO's Indian summer 145

economic model to registered births in 1912, 'It would seem therefore that the statement that the population is being recruited out of due proportion from its least successful and progressive elements receives confirmation from these figures.'[65] It was only in the 1920s that Stevenson appears to have fully accepted his model as an adequate tool of social analysis, and to have developed the novel concept of contraceptive diffusion to counter eugenic fears.[66] The GRO's *Annual report*, however, had begun to expand in the late 1890s[67], and many of the statistical innovations noted above, such as the reinstitution of the Superintendent's Letter and the introduction of new tables on infant mortality, were in place before Stevenson or Newsholme were even appointed to their central government posts in 1909.

In addition, one must question whether Stevenson himself always saw social class in terms of the detailed operation of certain impersonal and economic forces acting within and upon the country's poorer communities. Stevenson conceived of the effect of wealth, or the lack of it, on health not in terms of materialist life chances but as a product of 'culture', or the differential allocation of life skills. As he explained to the Royal Statistical Society in 1928:

> the lower mortality of the wealthier classes depends less upon wealth itself than upon culture, extending to matters of hygiene, generally on the whole associated with it. ... But culture is more easily estimated, as between occupations, than wealth, so the occupational basis of social grading has a wholesome tendency to emphasize it.[68]

It was left to Major Greenwood, who was chairing the session, to point out that in any community there must exist a group of people whose purchasing power was so small that no amount of culture in Stevenson's sense could possibly enable them to

---

[65] *75th ARRG for 1912*, p. xxviii.
[66] Szreter, *Fertility, class and gender in Britain*, pp. 254–71.
[67] Whilst the *Annual report* published in 1895 contained only 13 pages of text, that published in 1899 contained 44 pages: GRO, *57th ARRG for 1894*, pp. v–xviii; GRO, *60th ARRG for 1897*, pp. v–xlix.
[68] Stevenson, 'The vital statistics of wealth and poverty', p. 209.

provide adequately for the family unit. The problem then turned upon the definition of such a minimum standard, and the estimation of the numbers of persons who fell below it.[69]

Even after the First World War the GRO did not initially make very much use of its new model of socio-economic class. Stevenson indeed spent much time and ingenuity on analysing the returns to the 1911 fertility survey[70] but similar questions did not reappear in the decennial census until 1951.[71] After the discussion of the socio-economic class of fathers of babies born in 1912 in the *Annual report* for that year, births were not analysed in these terms again in the GRO's annual publications in the inter-war period.[72] There was a brief analysis in the *Statistical review* for 1937 of married women with whom a child under one was enumerated in the 1931 census broken down according to the socio-economic grouping of husbands, although this was not published until 1940.[73] A short, five page, discussion of legitimate births by fathers in socio-economic groups appeared in the *1921 Decennial supplement*, published in 1927, but the *Decennial supplement* volume relating to fertility for 1931 was not released until the 1950s.[74]

Similarly, Stevenson's model of socio-economic groupings was not used extensively by the GRO to study mortality differentials until the late 1930s. After discussing infant mortality by social class in five pages of the *Annual report* for 1911[75], Stevenson did not return to the subject in the GRO's annual publications. Instead, he concentrated on analysing mortality in terms of such traditional variables as age, locality, settlement type, and overcrowding. His work on class mortality gradients

---

[69] Ibid., p. 221.
[70] GRO, *75th ARRG for 1912*, pp. viii, xxii–xxxii; *Census 1911, Vol. XIII, Fertility of marriage report*, Part; *Census 1911, Vol. XIII, Fertility of marriage report*, Part 2; Stevenson, 'The fertility of various social classes'.
[71] Office of Population Censuses and Surveys and the General Register Office, Edinburgh, *Guide to census reports*, pp. 221–30.
[72] GRO, *75th ARRG for 1912*, pp. xxii–xxxii.
[73] GRO, *Registrar General's statistical review for 1937* (London, 1940), p. 208.
[74] GRO, *The Registra General's decennial supplement, England and Wales 1921*. Part II, pp. xcv–c; GRO, *The Registrar General's decennial supplement, England and Wales 1931*. Part IIb.
[75] GRO, *74th ARRG for 1911* (London, 1913), pp. xl–xlv.

was confined to analyses in *Decennial supplements* published in 1923 and 1927, and to a few articles commenting upon these.[76] The data on class mortality rates in the years 1910–12 which was released in 1923 almost failed to appear because Stevenson felt that it was now out of date. It was only 'strong representations' from outside the Office that had persuaded him to proceed with their publication.[77] After Stevenson's death in 1931 there were few allusions to the effects of social class on mortality rates in the GRO's annual *Statistical reviews* until that published in 1936. This contained specific data on maternal mortality by social class, and included the proportion of males over 14 in social classes IV and V in a regression equation showing the general effect of latitude, housing density and social class on mortality rates.[78] From 1938 onwards, however, mortality rates for specific diseases began to be studied in terms of social class.[79] The year 1938 also saw the publication of the volume of the *1931 Decennial supplement* dealing with occupational mortality that examined overall class mortality differentials, and with respect to specific diseases.[80] But even this modest level of social analysis petered out in the 1940s.

Thus, Stevenson's model of socio-economic groupings can hardly be said to have become a central, and regularly deployed, weapon in the GRO's statistical armoury until the late 1930s at the earliest. If poverty was discussed as a factor in explaining mortality, the key variables used by the GRO were various indices of overcrowding rather than social class. Indeed, as Charles Webster has argued, the relative dearth of official information on the effect of such variables on mortality hampered

---

[76] *Supplement to the 75th annual report of the Registrar General: Part IV Mortality of men in certain occupations in the three years 1910, 1911, and 1912* , pp. ix–xii, 3–6; *The Registrar General's decennial supplement, England and Wales 1921. Part II. Occupational mortality, fertility and infant mortality*, pp. v–lii; Stevenson, 'The social distribution of mortality from different causes in England and Wales'; Stevenson, 'The vital statistics of wealth and poverty'.

[77] GRO, *Supplement to the 75th annual report of the Registrar General: Part IV*, p. i.

[78] GRO, *Registrar General's statistical review for 1931*, pp. 26–8; GRO, *Registrar General's statistical review for 1934*, pp. 130–1, 150–5.

[79] GRO, *Registrar General's statistical review for 1936*, pp. 90–1, 101–2.

[80] GRO, *The Registrar General's decennial supplement, England and Wales 1931. Part IIa*, pp. 19–74.

public debate on the subject in the inter-war period.[81] In this he is merely repeating an observation made at the time by Richard Titmuss in his *Poverty and population* of 1938.[82] The concept of socio-economic groupings was, of course, used by medical statisticians outside the GRO in the inter-war period but often only in order to dismiss its usefulness in explaining mortality.[83]

Various reasons could be put forward for the general unwillingness to exploit Stevenson's methodological innovation of 1911. In part, the staff of the GRO were uncertain about the reliability and relevance of the socio-economic model, especially regarding its applicability to mortality analysis. Stevenson was concerned himself about the extent to which Socio-Economic Group V, the unskilled category, failed to pick up all labourers[84], and his doubts regarding the consistency of his model were shared by subsequent officers of the GRO.[85] There was also the general problem of the tendency of the lowest social classes to recruit those whose poor physical or mental health prevented them from holding down better jobs. The association of high relative mortality with low social status might merely reflect a 'recruitment effect' rather than true causation.[86] Also, the model of socio-economic groupings had been designed to show the effect of social status and position, seen as synonymous with levels of intelligence, on fertility. But such a model might have less explanatory power when applied to mortality, where income levels might be a more useful variable.[87] It was not at all clear whether Stevenson's model related to status groups laying claim to the prestige associated with

---

[81] Webster, 'Healthy or hungry thirties', p. 116.
[82] Titmuss, *Poverty and population*, p. 253.
[83] e.g. McKinlay, 'Infant mortality and economic status'.
[84] Stevenson, 'The social distribution of mortality', pp. 384–5.
[85] Glass, 'A note on the occupational grouping used in tabulating the 1939 births', p. 99.
[86] GRO, *Supplement to the 75th annual report of the Registrar General: Part IV*, p. v. Stevenson himself appears to have overcome any doubts on this score: Stevenson, 'The vital statistics of wealth and poverty', pp. 212–14. For a modern restatement of this argument, see: Stern, 'Social mobility'.
[87] This was certainly the view of Glass in 'A note on the occupational grouping used in tabulating the 1939 births', p. 98.

Eugenics and the GRO's Indian summer

certain social positions, or to economic classes with quantifiable purchasing power within the market.[88] Indeed, at least one modern medical statistician has queried whether the deployment of the model of socio-economic grouping in the analysis of differential mortality is really helpful, or whether it gives a spurious appearance of knowledge whilst obscuring the real determinants of inequality in terms of factors such as income, wealth, education, and housing tenure.[89] As will be discussed in Chapter 7, the lack of interest in social class might also have reflected the politics of health in inter-war Whitehall, and the subordination of the GRO to the interests of the Ministry of Health.

The Ministry wished to play down the impact of social class on mortality for political reasons. Thus, the brief effervescence of the GRO's interest in the impact of social class on fertility and mortality seems to have occurred at a period when Stevenson was under the direct influence of Sir Bernard Mallet, who was himself interested in eugenic ideas. This may be another reason for urging caution when attempting to see Stevenson's work in terms of an environmentalist repost to the eugenicists.

The New Liberalism and the expansion of the GRO's resources

The emphasis on the conflicts between eugenicists and public health environmentalists has also distracted attention from the timing of the expansion of the GRO's staffing resources upon which the statistical innovations of the Edwardian period were, in part, based. As already noted in Chapter 4, the easing of the staffing constraints on the GRO appears to have begun in 1899 with the removal of Brydges Hennicker's control over the Office's affairs. Again, this was nearly a decade before the

---

[88] For the distinction see, Eldridge (ed.), *Max Weber*, pp. 86–92.
[89] Wilkinson, 'Socio-economic differences in mortality', pp. 18–19. Wilkinson has since become more positive about the use of the socio-economic group model since he now sees the psychosocial effects of social inequality as being the main determinant of inequalities in health: Wilkinson, *Unhealthy societies. The afflictions of inequality*, p. 23.

appointment of either Stevenson to the GRO, or of Newsholme to the LGB. As can be seen from Table 3:1 and Figure 4:1, the improvement in the GRO's staffing continued in the Edwardian period, its complement increasing by nearly a third between 1895 and the financial year 1912/13. But little of this expansion appears to have been directly due to an intellectual struggle with eugenics.

A crucial innovation in the GRO's staffing structure was the replacement of men by women and boy clerks, which was a general feature of the British Civil Service in the early twentieth century.[90] One of the first acts of Sir Reginald MacLeod on replacing Henniker as Registrar General in January 1900 was to ask the Treasury for permission to employ four women typists to type out letters and certificates.[91] The addition of further female typists led to the appointment of a female clerk to check their work in 1902.[92] In the same year MacLeod also persuaded the Treasury to add four boy copyists to his complement to be 'employed in summarising the details as to ages and causes of death for tables,' which were to be included in the *Decennial supplement* for 1891 to 1901.[93] Sir William Dunbar, who succeeded MacLeod in 1902, continued this trend, replacing a male assistant clerk with a woman clerk in 1904 on a salary scale of £55 to £100.[94] At this date an assistant clerk could reach a maximum wage of £470, providing the Treasury with ample reasons for supporting the substitution.[95] By 1910 the GRO was employing nine female typists, five on correspondence and four on certificates. The new Registrar General, Sir Bernard Mallet, asked the Treasury to allow him to replace the four on certificates with five boy clerks producing them by hand, whilst moving the typists over to correspondence.[96] The next year saw

[90] Zimmeck, 'Strategies and stratagems for the employment of women'.
[91] PRO: RG 29/3, p. 266; RG 29/7, p. 163.
[92] PRO: RG 29/3, p. 273; RG 29/7, p. 167; RG 29/3, p. 285; RG 29/7, pp. 172–3.
[93] PRO: RG 29/3, p. 280; RG 29/7, p. 171.
[94] PRO: RG 29/3, p. 324; RG 29/7, p. 191.
[95] PRO: RG 29/3, insert between pages 279 and 280.
[96] PRO: RG 29/4, pp. 95–6. The Treasury only allowed him two boy clerks 'experimentally': PRO: RG 29/8, p. 49.

the addition of sixteen typists to the complement in order to produce the indexes of births, marriages and deaths which had previously been typeset by the Stationery Office, at a net saving to the Treasury of £2,480.[97]

As will be discussed shortly, new forms of social citizenship established in the Edwardian period also placed extra responsibilities upon the GRO. The introduction of old age pensions in 1909, and the need to verify ages for pension officers, led to the addition of two second division clerks, five assistant clerks, three boy clerks, and three porters to the Office's complement.[98] By the end of the following year, the old age pensions section of the Office comprised two second division clerks, 19 assistant clerks, and six boy clerks, and was roughly comparable in size to the Statistical Department (see Table 3:1). Even then, the Registrar General was asking the Treasury for permission to employ the section's entire staff on overtime because of the pressure of work.[99]

Little of the expansion discussed so far related directly to the GRO's statistical staff, although the gradual easing of general clerical constraints must have helped with the provision of supervisory grades during the analysis of the 1911 census data. But the Statistical Department also saw an expansion in this period as its complement increased from 19 in 1895 to 27 in 1905 (see Table 3:1). But even this extra staffing cannot be put down solely to the results of intellectual debates with the eugenics movement. When Dunbar put forward a claim to the Treasury in 1905 for more statistical clerks, he did indeed mention new, but unspecified, lines of statistical enquiry, and returns for other departments, as reasons for his request. But he also pointed to the increase in population; earlier publication of reports; changes to district boundaries; and the extra work associated with the 1896 London Equalisation of Rates Act and the 1898 Marriage Act, as justifying such increases.[100] In addition, the abolition of piecework in the Office in 1911, which reduced

[97] PRO: RG 29/4. p. 134; RG 29/8, p. 59.
[98] PRO: RG 29/8, p. 23.
[99] PRO: RG 29/4, pp. 107, 119; RG 29/8, pp. 51–2.
[100] PRO: RG 29/3, pp. 348–50.

both staff earnings and output, led to the Treasury sanctioning an increase of five assistant and two boy clerks in the Statistical Department.[101]

Dunbar's list of 'new' pressures on the Statistical Department was, however, one that Brydges Henniker could have made at any time in the 1890s. Similarly, the effects of the abolition of piecework were analogous to those consequent upon the reduction in the Office's working day in 1891 but on the earlier occasion the Treasury had not agreed to an increase in staff numbers. Why, therefore, was the Treasury more amenable to staff expansion in the Edwardian GRO than in the last decade of the reign of Queen Victoria? The replacement of Henniker by more dynamic Registrar Generals appears to have been a factor but the Office's staffing structure also needs to be put in the much broader context of the general growth of Whitehall. According to Finer (Table 4:1), the British Civil Service expanded by 140 per cent between 1901 and 1914, whilst the LGB saw its staffing complement more than double between 1902 and 1912, from 435 to 963.[102] This appears to represent a relaxation of Treasury control during a period of growing demands for state intervention in both economy and society in pursuit of 'national efficiency'. Central government was being expanded to help mobilise the nation in pursuit of national and imperial advantage.[103]

Civil registration and developing forms of citizenship in the early twentieth century

A discussion of the expanding role of the GRO in the early twentieth century would, however, be incomplete without considering the manner in which the registration system underpinned new social rights in the early twentieth century. The legislation of the Liberal administrations of 1905 to 1915 began

---

[101] PRO: RG 29/4, pp. 115–16, 173–8; RG 29/8, p. 74.
[102] MacLeod, 'The frustration of state medicine', p. 40; MacLeod, *Treasury control and social administration*, p. 48.
[103] Cohen, *The growth of the British Civil Service*, pp. 163–4; Roseveare, *The Treasury*, pp. 186, 199–200; MacLeod, *Treasury control and social administration*, p. 38; Davidson, *Whitehall and the labour problem*, pp. 169–77.

a shift towards the still closer central organisation of welfare provision, and the introduction of nationally funded benefits.[104] This shift also reflected the pursuit of national efficiency in an age of imperial rivalry, drawing upon the perceived successes of Bismarck's social policies in Imperial Germany. In addition, the Liberal Party was also under pressure from the political demands of expanding numbers of working-class voters to drop Gladstone's fiscal conservatism.[105] State policy and the interests of citizens were working in the same direction. The survey work of Charles Booth in London, and of Seebohm Rowntree in York, had also revealed the structural poverty associated with low pay, underemployment and old age, which could not be overcome by the strenuous personal activity of the idealised working-class citizen of the mid-Victorian period.[106]

In 1908 the Liberals introduced old age pensions in the Old Age Pensions Acts of that year. The pension was paid from 1 January 1909 to those over the age of 70, provided that they had not been imprisoned during the ten years preceding their claim; were not aliens; and could satisfy the pensions authority that they had not been guilty of 'habitual failure to work . . .'. In all, 490,000 individuals qualified for the new transfer income which was to be paid out of central taxes.[107] As already noted, the GRO undertook to verify the ages of persons claiming the pension for the Board of Customs and Excise. Similarly, the 1911 National Insurance Act set up a nationally administered health insurance scheme. Employers and employees were to pay contributions into approved societies, to which the Treasury added a smaller subvention. In return the insured could claim the services of a doctor, special medical provision in the event of contracting tuberculosis, and certain cash benefits – sickness and disability benefits. The scheme was administered by local

---

[104] For the impact of New Liberalism of social policy, see: Freeden, *The New Liberalism*; Collini, *Liberalism and sociology*.
[105] Freeden, *The New Liberalism*, pp. 25–75; Searle, *The quest for national efficiency*, pp. 1–106; Thane, *The foundations of the Welfare State*, pp. 4–47; Finlayson, *Citizens, State, and social welfare in Britain*, pp. 108–25, 151, 164.
[106] Booth, *Life and labour of the people*, Series 1, 2 vols.; Booth, *Life and labour of the people*, Series 2, 10 vols.; Booth, *Life and labour of the people*, Series 3, 17 vols.; Rowntree, *Poverty*.
[107] Thane, *The foundations of the Welfare State*, pp. 83–4.

insurance committees with a National Health Insurance Joint Committee operating under the Treasury at the centre.[108] To facilitate the workings of the scheme, any person seeking proof of age for the purposes of claiming benefits under the Act could obtain a birth certificate at a cheap concessionary rate.

The mass mobilisation of Britain during the First World War also saw the creation of new, centrally funded, rights to transfer payments from the tax-payer to special categories of citizens. The war years, for example, saw a flood of searches in the GRO to assist dependants of men serving at the Front to obtain evidence of marriage and of the birth of children under 16 years of age, in connection with claims for army and naval separation allowances.[109] Many of these had, of course, to be converted to war widow's pensions by 1918. The Great War also saw the mass conscription of men and women into the armed forces, munitions industries and economy in general. As will be explained in Chapter 7, the GRO was at the heart of the system of national registration which underpinned these novel social obligations.

The introduction of new rights organised at the level of the nation state added greatly to the flow of work within the GRO. Between 1900 and 1920 the number of searches in the register paid for by fees increased from 57,895 per annum to 134,747.[110] The total number of searches overall was far higher at 284,000 in 1920, with 108,000 certificates issued.[111] A large proportion of this growth represented old age pension applications. These began to reach the Office in October 1908, and by the end of December it had received notification of 20,855 claims.[112] By 1913 the number of such searches had risen to over 70,000 per annum.[113] In the immediate aftermath of the Great War, the search rooms at the GRO were receiving over 200 requests per day for verification of ages from pensions officers, and between 10 and 20 per day from the Appeals Branch

---

[108] Gilbert, *The evolution of National Insurance*.
[109] GRO, *77th ARRG for 1914*, p. viii; Thane, *The foundations of the Welfare State*, pp. 127–8.
[110] GRO, *Registrar General's statistical review for 1950:* Civil tables, p. 80.
[111] GRO, *Registrar General's statistical review for 1921*, p. 115.
[112] GRO, *72nd ARRG for 1909*, p. cxxviii.
[113] GRO, *76th ARRG for 1913*, p. lxxxv.

of the Ministry of Health, pensions committees and the general public. At the same date, the 'War Section' was said to have been completing between 2,000 and 3,000 searches per week, 'in connection with liability for service in the Army under the Military Services Acts, claims for separation allowances, pensions, special allowances, service in the State Munitions Establishments, liability of young soldiers for service abroad, etc.'.[114]

Such developments can plainly be seen as either 'progressive', or as a strategy of amelioration to help defuse opposition to a society based on class antagonism. They can also be seen as the replacement of local citizenship based on claims on the parish of birth or of settlement under the Poor Law, with a new citizenship based on national rights and obligations. Whatever their origins, they led to the creation of a new type of state – one that was more responsive to democratic pressure, and was more expansive as a result. The GRO inevitably shared in this expansion. But before pursuing these themes further in the twentieth century, some consideration needs to be given to the important changes introduced in the Office's technologies for statistical production in the Edwardian period.

---

[114] PRO: RG 20/76 Search Room: organisation and staffing 1919–20.

6

State medical statistics, the dawn of computing, and the Edwardian mathematical revolution

Intellectual and technological explanations of history

Historians have a tendency to suppose that intellectual debate and the clash of ideas are the key motive forces in human affairs. This may be a comforting thought for intellectuals, and may be true on certain occasions, but as an explanation of overall change it appears somewhat lacking. As well as studying the intellectual struggles between eugenicists and public health environmentalists, it might also be productive to examine the impact the former, and some of their associates, had on statistical manipulation. This needs to be conceived not in terms of a conscious contest of ideas but of the consequential effects of the adoption of new techniques.

When considering the development of statistical production with the GRO, much attention has been given to the intellectual history of the institution. However, an equally powerful argument can be made that the key determinant of the form of statistics created by the Victorian GRO was its need to create simple representations of data that did not overburden its limited resources and cumbersome manual methods of data manipulation. These have been mentioned in passing above, and covered by the present author in more detail elsewhere.[1] However, eugenics helped to trigger a revolution in data processing technologies within the GRO because of the knock-

---

[1] Higgs, 'The linguistic construction of social and medical categories'; Higgs, 'The General Register Office and the tabulation of data'. For an earlier version of some of aspects of the discussion in this chapter, see: Higgs, 'The Statistical Big Bang of 1911'.

on effects of the introduction of the fertility survey into the 1911 census. The increasing complexity of data, and of the forms of analysis applied to them, led to the adoption of machine tabulation. The new method of working had the effect of expanding the GRO's capacity for manipulating information, and initiated a split between statistical analysis and the work of data preparation. At the same time, the new statistical methods pioneered by Galton and Pearson, and the development of a cadre of statisticians adept in them, helped to undermine the professional standing of the GRO in Whitehall.

None of this should be seen, however, in terms of technological determinism. Machines, or new techniques, are only adopted within organisations because they fulfil a need, or solve a problem. But the recognition that a problem exists, or that an organisation has a particular need, is not an inevitability. Indeed, much of history could be conceived of in terms of the failures of individuals, organisation, or states, to achieve such a level of awareness. Technology, therefore, is not of itself the key to understanding innovation. The latter has to be understood as a social, organisational and historical process in its own right. To ignore this is to impoverish the understanding of technology as the application of scientific principles to meet social demands. The history of technology is, after all, littered with examples of good ideas that lacked an application until a happy confluence of circumstances led to their adoption and implementation.

Data processing technology and the politics of innovation

In order to analyse this process within the GRO, the present work will draw upon the useful work of Everett Rogers, an American management theorist.[2] Although much management and organisation theory emanating from the USA is made up of a few highly formalised platitudes drawn out at great length, it is sometimes useful to spell out the obvious, since this process highlights the necessary stages in an argument. Rogers notes

---

[2] Rogers, *The diffusion of innovations*.

that the diffusion of an innovation requires four elements – (1) the technical innovation itself, which (2) is communicated through certain channels, (3) over time, and (4) amongst the members of a social system.[3] The latter element is plainly crucial – technology has to be taken up within a social setting, and Rogers stresses the importance of the norms of organisations and the crucial role of 'opinion leaders' in this process.[4] The take-up of technical innovations within organisations is not inevitable, and Rogers posits various stages in the innovation process within them:

1) agenda-setting – a 'performance gap' opens up 'between how an organisation's members perceive its performance, in comparison to what they feel it should be . . .';
2) matching – can a communicated innovation solve the problem? – a trial is undertaken;
3) redefining/restructuring of the innovation to meet the organisation's needs;
4) extrapolating uses of the innovation throughout the organisation;
5) routinisation.

Rogers also points to 'organisational slack' – the degree to which uncommitted resources are available to an organisation – as an important factor in the process of innovation.[5] The key assumption underlying Rogers' model is that change or innovation within organisations is not automatic. Organisations can often spend much of their time seeking ways of avoiding changes, which cause uncertainty, and upset power relations within bureaucratic systems. In many ways modern bureaucracies, whether state or capitalist, have been established with the very purpose of removing room for doubt or debate, which change threatens. The organisational tendency to inertia may thus oppose and defeat innovation.

[3] Ibid., p. 10.
[4] Ibid., pp. 26–7.
[5] Ibid., p. 361.

State medical statistics in the Edwardian era 159

The GRO has certainly been accused of inertia with respect to the delay in introducing machine tabulation of data into the Census Office until 1911.[6] Mechanical means of processing census data had been invented by the American Herman Hollerith over 20 years before for use in the US enumeration of 1890. Hollerith patented two machines, one that punched holes at predetermined positions in cards of eighty columns, each hole standing for a numerical value. In this manner information could be stored on cards in a quantitative form. Hollerith's other machine, which read these cards, had what was known as a 'pin-box' containing spring-loaded needles, one for each possible hole in a card. The pin-box was brought down over each card in turn, and those needles that met an unpunched surface were depressed. Those that passed through a hole made an electrical contact below, and the resulting passage of current flowed, by a series of relays, to dials which then moved forward one position. Thus one could count the value or occurrence of a variable, or the co-incidence of two or more, at great speed.[7]

By breaking data-processing down into, first, coding and punching, and then, secondly, into data manipulation, mechanical processing greatly increased computing potential. In the tabling/ticking system used in the Victorian GRO one identified the value of a variable in the process of manipulating the data, by placing a tick in a particular box in a ticking sheet. Each set of data had to be abstracted using a new tabling sheet. In the mechanical method, the intellectual work of data manipulation was separated from mechanised coding via punching. The latter only needed to be performed once, whilst the resulting encoded data could be manipulated automatically at great speed as many times, and for as many different combination of variables, as was desired. As Sir Bernard Mallet explained to the Royal Statistical Society in 1916:

---

[6] Szreter, 'The GRO and the public health movement in Britain', p. 462, n. 96; Campbell-Kelly, *ICL*, p. 28.
[7] Merriam, 'The evolution of American census-taking'; Campbell-Kelly, *ICL*, pp. 8–13; Cortada, *Before the computer*, pp. 49–63; Heide, 'Shaping technology'.

Once the labour of preparing the cards required for the routine tabulation as previously carried out has been accomplished, it becomes a very simple matter to obtain records of additional combinations of the facts recorded, whereas under the system previously employed each additional tabulation had to be undertaken independently, the record of one combination of facts not contributing in any way to the preparation of that of another.[8]

The range and complexity of questions to be asked of survey data could thus be greatly increased. This represented, of course, the invention of the modern database.

But if the new system was so powerful, why did the GRO wait till 1911 to introduce it? Hollerith had developed his machinery in the 1880s, and had written to the 1890 Treasury Committee on the Census extolling its use.[9] The Committee had duly recommended that the use of the new technology should be investigated by the GRO.[10] Hollerith backed this up by reading a paper before the Royal Statistical Society in London in 1894, in which he described the workings of his inventions.[11] This was one of a number of visits Hollerith made to Europe in the 1890s to advertise his equipment, which was taken up in Austria and Norway, and its use actively explored in Russia and France.[12] Thus, the first two elements in Rogers' schema for successful innovation – invention and dissemination of information – had been put in place.

The lag in introducing the equipment into England and Wales might possibly be put down to bad timing. Until the passing of the 1920 Census Act, expenditure on census taking was only authorised on a temporary basis every ten years, and the whole central census apparatus had to be created from scratch on every occasion. This meant that experiments with mechanical tabulation had to be carried out in the immediate run-up to

---

[8] Mallet, 'The organisation of registration', p. 8.
[9] *Report of the 1890 Treasury Committee on the Census*, Appendix 9.
[10] Ibid., p. xi.
[11] Hollerith, 'The electrical tabulating machine'.
[12] Campbell-Kelly, *ICL*, p. 13.

the decennial enumeration. Information about Hollerith's untried machinery probably came too late for its use in the British census of 1891. In the years immediately preceding the 1901 census, on the other hand, the then Registrar General, Brydges Henniker, was ill and nearing retirement, and may have been incapable of steering the Office on a new course.[13] It was not until the lead in to the 1911 enumeration, therefore, that the new system of working could be tested, and the technology installed.

T. H. C. Stevenson, however, saw the introduction of machine tabulation as a triumph over the forces of re-action within the GRO. As he wrote to Mallet, in 1920:

> It was assumed by those whose statistical experience and proved ability entitled their opinion to most respect that our work was too complicated to be suited to machine tabulation. Fortunately you and I, in our then condition of inexperience but perhaps greater receptivity, formed the opposite opinion....[14]

This brings to the fore the role of the institutional culture of the Office in the late Victorian and Edwardian periods. As one can see from the readiness with which the GRO introduced mechanical means of calculation into its work, the failure to implement new methods of mechanical data processing cannot be put down solely to an abhorrence of new technology.[15] The GRO, however, was profoundly hostile to any innovation in the scope and complexity of the questions to be asked of the public. A long-standing belief in the inadequacies of the agents involved in census-taking, and the limited purposes for which state data gathering was considered permissible, precluded any expansion in the size or complexity of the census schedule.[16] One of George Graham's final acts as Registrar General had

---

[13] On Henniker's illness see a letter of the GRO's chief clerk to the Treasury of April 1899: PRO: RG 29/3, p. 226.
[14] PRO: MH 78/114 GRO co-ordination of work with Ministry of Health and appointment of Registrar General: 'Note by Dr Stevenson' dated 25 Jan. 1920.
[15] Higgs, 'The General Register Office and the tabulation of data', pp. 222–5.
[16] Higgs, 'The General Register Office and the tabulation of data'.

been to inform the Treasury wearily that he believed with respect to the 1881 census that:

> learned societies, statists, the Social Science Association, the Statistical Society, the Nonconformists etc. will urge Her Majesty's Government to obtain in addition to what were enquired into in 1871, particulars as to religion, health of the people, an industrial census as to wages, etc., the colour of hair of different persons, etc., etc.[17]

The implication was plainly that such demands were essentially frivolous and should be resisted. As late as 1908, Archer Bellingham, the GRO's superintendent of records, could still claim in an official memorandum on the census, that:

> it is better to ask for a moderate amount of information, and to concentrate efforts on getting that amount with the greatest attainable accuracy than to call for much information on many subjects even though they may appear important to certain sections of the public.[18]

Also, and following Roger's terminology, one should stress the importance of the role of the main 'opinion former' within the GRO, the Registrar General. Henniker was hardly the man to pioneer such a fundamental innovation, or to put pressure on the Treasury for extra expenditure on preparations for the census. The increasing workload of the GRO, the inevitable consequence of population growth and a fixed staff complement in the last decade of the nineteenth century, may also have hindered the introduction of new technologies. There was precious little 'slack' to plan and experiment with new forms of data processing. The organisational culture of the GRO was not, therefore, conducive to change.

But this reluctance to expand the number of questions on the census schedule was now overcome by Mallet's and Stevenson's desire to use the 1911 enumeration to explore the issue of

[17] PRO: RG 29/2, p. 282.
[18] PRO: RG 19/45, p. 42.

differential class fertility that had been thrown up by the eugenics movement. The fundamental eugenic attack on the Victorian concept of the relationship between citizen and state was moving the latter towards greater intervention. At the same time, the increasingly interventionist activities of other central government departments were also pushing the GRO in the same direction. The 1901 census had seen the introduction of a question on home working at the bequest of the Home Office, and the 1911 enumeration saw the introduction of a column on the schedule asking for information on the industry in which individuals worked in addition to their personal occupation. As discussed above, this innovation had been urged on the GRO in an interdepartmental committee on the census by a representative of the Board of Trade. The Board was interested in establishing how many people worked in particular industries and thus how many were dependent upon them – they wanted to know more about how many bricklayers were directly engaged by cotton mill owners, for example, and not just the total size of the bricklaying population. This was plainly aimed at gaining more data on the effects of a depression in one trade on levels of unemployment.[19]

The staff of the GRO were aware that such changes to the census schedule and the increasing complexity of data analysis they implied would overburden the 'ticking' system of data processing. According to Archer Bellingham, the GRO's superintendent of records, in his 1908 memorandum on the census:

> The ticking system hitherto in use in this Office has been found to combine economy with reasonable accuracy, and has therefore been regarded as suitable for the kind of work required. . . . Moreover, the comparatively simple nature of the returns (except in the case of occupations) and the absence of any attempt at a degree of analysis incompatible with the limits of a single abstract sheet, have prevented the necessary limitations of the ticking system making themselves

---

[19] PRO: RG 19/48B Dept committee on the census of 1911 (Oct. 1909–Apr. 1911), p. 103; Davidson, *Whitehall and the labour problem*, pp. 139–43.

inconveniently felt. If, however, any alteration in the form of the returns, any considerable increase in the scope of the Census inquiry, or any greater detail in the presentation of results were required, it might be found that the ticking system would be inadequate.[20]

This eventuality had now come to pass, and Mallet had to explain to the Treasury that the fertility analysis on the 1911 data could not be performed by the ticking method, and that on occupation and industry, only with the greatest difficulty.[21] The ability of Hollerith's inventions to analyse data in new, complex ways was also stressed in the contemporary technical press.[22] Mallet assured the Treasury, however, that the cost of purchasing machine tabulators would be counterbalanced by the shortening of the time require for data processing, and by the replacement of the labour of clerks by the employment of young girls as machine punchers on low wages.[23] Such girls, 170 in number and aged between 14 and 15, were to be employed straight from school. It was hoped that the work would, 'attract a somewhat higher type than the factory girl, and need not hinder her from getting permanent employment afterwards.'[24]

In practice, however, the new method of working did not save money. Whilst the 1901 census cost 20 per cent more than that of 1891, that of 1911 was 50 per cent more expensive to process than its predecessor.[25] Mallet fell back, therefore, on stressing the impossibility of undertaking the new forms of research under the ticking system, and the maintenance of the unit cost of statistical production. As he explained to the Treasury:

[20] PRO: RG 19/45, p. 42. See also, PRO: RG 19/186 Memo on census taking in Austria, Hungary, England and Wales, France, German Empire, the Netherlands, Switzerland, USA: section on England and Wales.
[21] PRO: STAT 12/10/9, letter of 25 October 1910; PRO: RG 29/4, pp. 108–9.
[22] Anon., 'The tabulator. No 1'; Anon., 'The tabulator. No 2'. See also Campbell-Kelly, *ICL*, p. 11.
[23] PRO: STAT 12/10/9, letter of 25 October 1910.
[24] PRO: RG 19/48B Departmental committee on the census of 1911 (Oct. 1909–Apr 1911), p. 181; RG 29/4, p. 123.
[25] PRO: RG 29/8, p. 69.

Due weight should also be given to the great increase both in the volume and complexity of the tabulation generally which has been caused by the necessity of satisfying the demands of the Home Office, the LGB, and the Board of Trade. It is now at all events clear, in my opinion, that the old methods would not have sufficed to deal with the material furnished in answers to the questions in the census schedule, and that even if the attempt had been made to retain those methods, the cost would have been quite as large as the new system.[26]

It is indicative of the GRO's new relationship with the Treasury that the Registrar General's excuses met with a sympathetic response from the latter, which recognised the 'difficulty of framing a reliable estimate owing to the adoption for the first time of the card system . . .'.[27]

Thus, by the mid-1900s all the elements were in place for the introduction of technological innovation. A gap was plainly perceived between what the GRO wanted to do, or what other central departments required, and what its current data-processing technologies would allow it to do. The factors that had encouraged a negative and defensive ethos within the Office had been removed, whilst the leading figures in the organisation were predisposed to change. In addition, Roger's requirement of institutional 'slack' was now in place given the rapid expansion of the GRO's staff in the Edwardian period. Knowledge previously accumulated was now acted upon. In addition, the willingness to demand more information from members of the public may also have reflected a greater confidence in their ability to provide reliable information. The *1901 census report* contained a number of allusions to the improvement in the quality of responses to census questions, and the GRO appears to have become rather more sanguine with regard to levels of functional literacy.[28] This, arguably, reflected the effects of compulsory elementary education, introduced in England and

[26] PRO: RG 29/4, p. 184.
[27] PRO: RG 29/8, p. 69.
[28] *1901 census report*, pp. 74, 90.

Wales in in the late Victorian period, feeding through into literacy levels. Thus, state intervention, and the imposition of new duties on citizens, were increasing both the demand for reliable social data and the potential for its supply.

The Victorian GRO's desire to limit the flows of information and the complexity of data analysis played a crucial role in retarding its take-up of technical innovation, and this can be seen as a significant example of the comparative retardation of the market for office technology in the UK in general. As Cortada has argued, it was the increasing flow of information in US businesses that led to the development of the modern computer industry, beginning with the production of adding and calculating machines in the nineteenth century.[29] He has also stressed rapid growth in GNP, business concentration and the growing complexity of business processes as the preconditions for the success of commercial computing in the USA since the 1950s.[30] Possibly the relatively smaller scale of British firms prevented the early development of a requirement for the information processing capacity required by US and German companies.[31] Certainly the UK was behind both the USA and Germany in developing office technology industries. In 1913, US manufacturers of typewriters exported products valued at over $11 million, or approximately five times the office products sent to the US from Germany, and 60 times those from the UK. In 1919, no British manufacturer of typewriters, cash registers, adding or mimeograph machines was amongst the top 200 corporations, but several US firms of this type were in the top tier.[32] Even in British industries that undertook vast data-processing operations, the use of card tabulators was not common before the First World War. The Prudential Insurance Company, for example, only introduced card-punch technology in 1915, and this was much in advance of the British banking

---

[29] Cortada, *Before the computer*, p. 7.
[30] Cortada, 'Economic preconditions that made possible application of commercial computing'.
[31] Yates, *Control through communication*, p. 271; Hannah, *The rise of the corporate economy*, pp. 24–8; Hannah and Kay, *Concentration in modern industry*, pp. 2–4.
[32] Cortada, *Before the computer*, pp. 38, 42.

sector.[33] The Prudential's innovation was directly related to the increased work associated with the administration of health insurance under the 1911 National Insurance Act. Again, new national rights and obligations were impacting on data processing methods.

The immaturity of the British office technology industry is amply illustrated in the case of the introduction of machine tabulation into the 1911 census. The Hollerith machines could not be imported direct from the USA, and a version of them had to be produced in England by the British Tabulating Machine Company (BTMC), a subsidiary of Hollerith's parent company established in 1904. In 1910 the BTMC only had four other customers for automatic tabulators in the whole UK, and the GRO contract, the first in government, was so large that it exceeded the company's total business to date. To finance the contract a bank loan of £3,000 had to be arranged on the personal security of the directors and their friends.[34] Everard Greene, the first general manager of the BTMC, described the company's agreement to undertake the 1911 census work as, 'a piece of cold-blooded cheek, in view of the magnitude of the census'.[35] The BTMC eventually became the British computer firm ICL but the British information market was never sufficient to allow it to compete with the emerging US giants such as IBM, which, in turn, had its origins in Hollerith's original US company.[36] In the British context, the introduction of machine tabulation into the census of 1911 was, therefore, a leap in the dark.

The association of the increasing complexity of data, and its analysis, and the introduction of mechanical means of data processing was not confined to England and Wales. The original introduction of machine tabulation into the US census was associated with the far greater range of information being sought by the American authorities in the late nineteenth century. This, in turn, reflected the demands by US congressmen and senators for information pertinent to the well-being of

[33] Campbell-Kelly, 'Large-scale data processing in the Prudential'.
[34] Campbell-Kelly, *ICL,* pp. 27–9.
[35] Greene, *The Beginnings,* p. 13.
[36] Campbell-Kelly, *ICL,* p. 22.

their constituents in the world's first mass democracy.[37] In his 1894 paper before the Royal Statistical Society, Hollerith stressed the role that increasing complexity played in the adoption of machine tabulation in the USA, noting that, 'Almost in every direction could be seen the need for combined or correlated statistics.'[38] This was also the view of the director of the US census, W. R. Merriam.[39] The census in late-Victorian England and Wales only asked for information on an individual's name, age, sex, marital status, relationship to head of household, occupation, employment status, birthplace, and medical disabilities. The 1910 US census, on the other hand, asked for name; relation to head; colour; sex; age; marital status; number of years in present marriage; number of children born to a mother and how many still living; birthplace; birthplace of both parents; year of immigration to the USA; whether naturalised or alien; whether able to speak English or other language spoken; occupation; industry in which occupied; whether employer or employed; whether in employment; number of weeks out of employment in 1909; whether able to read and write; whether a pupil at school; whether a house owner or tenant; if a veteran of Civil War; and whether deaf and dumb or blind.[40]

As in England and Wales, the introduction of machine tabulation elsewhere in Europe was also linked to an increase in the complexity of questions to be asked in national enumerations. The Registrar General for Scotland introduced mechanical tabulation into the 1911 enumeration of the northern kingdom at the same time as the new census enquiries introduced in England and Wales.[41] In France the adoption of a form of machine tabulation in 1896 was associated with the introduction into the census of new questions on employment and

---

[37] Anderson, *The American census*. See also, Beniger, *The control revolution*, pp. 390–425.

[38] Hollerith, 'The electrical tabulating machine', p. 678.

[39] Merriam, 'The evolution of American census-taking', p. 839.

[40] PRO: RG 19/186: Memorandum on census taking in Austria, Hungary, England and Wales, France, German Empire, the Netherlands, Switzerland, USA [1911]. For an example of the 1910 federal census schedule, see: Ruggles, Sobek and Gardner, 'Distributing large historical census samples', pp. 148–9.

[41] Scottish Record Office: GRO 5/1196 Statistical methods used in preparation of the Registrar General's reports, pp. 1–2.

economic operations. It was hoped that these would facilitate understanding of the contemporary economic crisis in France, which had led to unemployment and social tensions.[42] Similarly, van den Ende has argued that the introduction of such methods into the Dutch census in 1930 was directly related to the increasing complexity of the analysis involved.[43] This he ascribes to the expansion of government activity in the Netherlands, and to an extension of the participation of various social groups in public decision-making.[44] The factors leading to greater complexity in data analysis were plainly complex but the development across Europe of state-led strategies of social amelioration, national efficiency, and imperial aggrandisement, and the reaction to democratic aspirations, were plainly crucial.

The conclusions to be drawn from this analysis of the introduction of machine tabulation into the GRO are simple but have some interesting implications for the study of technological innovation. By itself research and invention do not lead to practical innovation, however widely disseminated the results of these activities are. Intellectuals and scientists seldom have the resources to turn their ideas into reality. The spread of technical innovations depends, therefore, on their implementation by commercial and state organisations. The latter process depends ultimately on the needs of those organisations and upon the extent to which their cultures are predisposed to change and innovation. This is why wars have always been periods associated with technical 'progress'. In the case of the GRO, a negative organisational culture changed into one predisposed to innovation because of broad shifts in the role of the institution, and because of specific changes in personnel. A gap had opened up between expectation and performance that machine tabulation could fill. The timing of change was determined by the perception that it was time for change.

---

[42] Desrosieres, 'Official statistics and medicine', pp. 525–6.
[43] Ende, 'The number factory', p. 18.
[44] See also, Snellen, 'From societal scanning to policy feedback', pp. 200–1; Ketelaar, 'Recordkeeping and office technology', pp. 215–17.

The consequences of technological innovation – extrapolation

Once the decision had been taken to introduce machine tabulation, the other elements in Rogers schema for innovation – redefinition, extrapolation, and routinisation – came into play. The Hollerith machinery, or at least the version of it available from the BTMC, was not ideally suited to the Office's specific needs, and had to be adapted.[45] Once its use had been perfected for census work, however, the GRO moved to use the new technology throughout the organisation.

The demographic and sanitary statistics published by the GRO in the *Annual reports* had two main failings in the late Victorian and Edwardian periods. Whilst the sanitary and public health administration of the country was based on urban and rural sanitary districts, data on births, marriages and deaths from the registration system was collected and published in registration districts. This meant that deaths per thousand, the basic measure of health, could not easily be calculated for the districts for which MOHs were responsible. In addition, deaths in institutions such as hospitals were not consistently transferred to place of usual residence. This increased the mortality rate for districts with such institutions, and decreased those elsewhere.[46] The first issue appears the more important today but it was the second that caused the most difficulties for the GRO in the early twentieth century.

Controversy over the districts for which mortality statistics were given in the *Annual reports* actually predated the establishment of sanitary districts in 1872.[47] However, the mismatch between registration districts and sanitary districts became more obvious when the GRO and the medical wing of the Privy Council were brought together in the LGB. One arm of the Board was producing statistics in a form that was not compatible with the needs of another. In the *Decennial supplement* for

---

[45] Greene, *The Beginnings*, p. 13.
[46] Mooney, Luckin and Tanner, 'Patient pathways'.
[47] Eyler, 'Mortality statistics and Victorian health policy', p. 350; Szreter, 'The GRO and the public health movement in Britain', pp. 461–2.

1881–90, published in 1895, John Tatham expressed his regret that the elaborate tables contained therein, 'cannot, except in a few instances, be utilised for definite administrative purposes because of the overlapping and confusion of areas which still frequently persist.'[48] In 1903 the County Councils Association called for the registration and sanitary districts to be unified, and the following year senior officers of the GRO agreed in discussions with the chief medical officer of the LGB that something needed to be done.[49]

There were, however, certain administrative and financial obstacles to the attainment of the desired objective. Either the GRO had to collect cause of death data on the basis of sanitary districts, or it had to continue the collection of data by registration districts and then abstract it according to the new boundaries. The first would mean new legislation to change the form of the local registration framework. As noted in Chapter 1, under the 1836 Registration and Marriage Acts, the poor law unions had been the territorial units upon which the registration districts had been based. The local registrars were appointed by the local Poor Law authorities, and were often local Poor Law officers. Although supervised and inspected by the GRO, they were paid fees for each certificate issued and the provision of other services, and were thus extremely independent. Reorganising the registration service into sanitary districts was a threat to the position of these men. As the Registrar General was aware, the formation of new districts would require compensation for those registrars whose jobs disappeared, or whose fees were reduced. It also meant ensuring that the sanitary authorities were willing to pay the registrars' fees and pensions.[50] Given the somewhat precarious influence that the GRO exerted over the registrars and local authorities, the Office was chary of undermining their organisation.

---

[48] GRO, *Supplement to the 55th annual report of the Registrar General*, p. v. He repeated his apologise 12 years later: GRO, *Supplement to the 65th annual report of the Registrar General*, p. lxiv.
[49] PRO: RG 29/22, pp 3, 28.
[50] GRO, *71st ARRG for 1908*, pp vi–vii; PRO: RG 48/1389 1912–1914 Discussions on proposals for a new Registration Act: minutes of the meeting of 19 August 1912.

Challenging the existing arrangements would also involve the GRO in the contentious issues of Poor Law and local government reform.[51]

The alternative solution – collecting cause of death data by registration district but abstracting it internally within the GRO according to sanitary districts – had serious financial implications. Transferring vital events from one set of areas to another under a system of manual data processing would require extra staff, as would the production of tables abstracted by both sets of administrative units in order to maintain the long-run comparability of time series. As the Registrar General commented in the *Annual report* for 1908, 'the change could not be introduced without some increase of staff and expenditure, and that this Department has, therefore, not a free hand in the matter.'[52] This was a not too oblique reference of the reluctance of the Treasury to pay for the extra clerical labour required. The answer to this dilemma was to use the new techniques of mechanical data processing which the GRO had perfected for the 1911 census, and which allowed the punching and analysis of deaths according to both registration and sanitary districts.[53] As the Registrar General explained in his *Annual report* for 1909, the new procedures involved in the shift from registration to sanitary districts would have thrown:

> so much additional work upon the Statistical Branch of this Department that they could hardly, for financial and other reasons, have been practicable without some modification of existing statistical methods. It has therefore been decided to introduce the use of cards in tabulating the returns of deaths. The need for re-arrangement of these returns in order to publish them under administrative areas has made the use of a card system necessary. Fortunately, objections to it on the score of expense has been simultaneously removed by the adaptation to the requirements of vital statistics of the system

---

[51] P G 48/1388 1913–14 Registration Reform Committee: Minutes, etc.
[52] GRO, *74th ARRG for 1911*, p. viii; GRO, *71st ARRG for 1908*, p. lxxv.
[53] Szreter, 'The GRO and the public health movement in Britain', p. 462.

of electrical sorting and counting of cards for the purpose of the approaching census.[54]

But this change to the geographical units used for data abstraction raised another problem — the treatment of deaths in institutions. The difficulties associated with institutional deaths had long been recognised by the GRO. Farr was commenting on the problems thrown up by the situation of large hospitals and lunatic asylums in certain registration districts as early as 1864.[55] Similarly, the *Annual report* for 1873 noted that, 'A considerable proportion of the deaths in the metropolitan hospitals occur among patients admitted to those institutions from districts situated outside the registration boundaries of London'.[56] In his *Report* for 1884, Brydges Henniker again argued that London had a high number of deaths from certain causes due to it 'being a place to which numerous persons betake themselves when surgical operations are required . . .'.[57] From 1869 onwards the *Annual reports* had contained a table of deaths in major institutions so that the public could make compensatory adjustments to local mortality rates.[58] The increase in the number of serious illnesses treated in hospitals in the nineteenth and early twentieth centuries, meant that more people died there, and so the more serious this problem became. In 1880, for example, deaths in institutions represented approximately nine per cent of all deaths in England and Wales but this had risen to 14 per cent by the turn of the century.[59] Similar statistical effects could be found in other countries.[60]

There is evidence that from the 1890s the Office routinely redistributed deaths in institutions in London back to place of residence. According to correspondence between the GRO and

[54] GRO, *72nd ARRG for 1909*, p. viii.
[55] GRO, *Supplement to the 25th annual report of the Registrar General*, pp. iv, xix.
[56] GRO, *36th ARRG for 1873*, p. liv.
[57] GRO, *47th ARRG for 1884*, p. xx.
[58] GRO, *32nd ARRG for 1869*, p. 195.
[59] GRO, *43rd ARRG for 1880*, p. 99; PRO: RG 29/23, p. 131; Lukin, 'Death and survival in the city'; Hardy, 'Death is the end of all disease', pp. 473, 480–1. See also, Eyler, 'Mortality statistics and Victorian health policy', p. 350.
[60] Emery, *Facts of life*, pp. 57–60, 137–54. For the problems created by this in London, see: Mooney, Luckin and Tanner, 'Patient pathways'.

the LGB, in 1890 the Registrar General decided to redistribute deaths occurring in the metropolitan asylums at Caterham, Leavesden and Darenth, and registrars of the relevant sub-districts were to furnish the GRO with weekly returns of such mortality.[61] In 1895, one of the Statistical Department's clerks was said in Treasury files to be occupied in 'Distributing deaths occurring in metropolitan hospitals etc. to the sanitary districts to which the deceased belonged ...'.[62] The crude death rates in the tables in the *Decennial supplement* for 1891 to 1900, published in 1907, were 'approximately adjusted for deaths in institutions'.[63] In his *Annual report* for 1908 the Registrar General maintained that in addition to the regular redistribution of deaths in London institutions, the 'mortality statistics relating to 75 other large towns and 142 smaller towns, as published in the quarterly returns and in the annual summary have been approximately corrected by the aid of special returns furnished to the Registrar General by the local registrars.'[64] Although this was plainly not a complete solution to the problem, it would have gone some way to rectifying distortions in the published data.

The introduction of a general system for redistributing institutional deaths in 1911 followed on from the decision to publish cause of death data by sanitary districts. For many years local registrars had sent information on mortality within their districts to the local MOHs, as well as sending copies of the certificates to the GRO. The MOHs used this data to construct tables for their own reports on mortality in their districts which were forwarded to the LGB. The latter had attempted for many years to encourage MOHs to redistribute deaths in hospitals to places of residence by 'swapping' deaths amongst themselves but this initiative had met with only limited success. MOHs 'exporting' deaths were quite happy to do so but those expected to 'import' them, thus increasing the recorded levels of mortality in their

---

[61] PRO: MH 19/195: Correspondence between the GRO and LGB, 27 August 1890.
[62] PRO: T 1/8954A/13154.
[63] GRO, *Supplement to the 65th annual report of the Registrar General*, p. ix.
[64] GRO, *71st ARRG for 1908*, p. vii.

districts, were naturally less co-operative. The vague addresses frequently found on certificates, meant that there was ample scope for MOHs to refuse to accept 'incoming' deaths. If the GRO failed to consistently transfer deaths to place of residence when it began to publish mortality data on the same basis as the MOHs, 'undesirable discrepancies' would appear between the two sets of medical data. This would undermine the public's faith in the accuracy of the whole statistical edifice on which the GRO had invested so much time and effort.[65] As a consequence, the Office was forced to take over the whole process of redistribution in order to organise it on a consistent basis. Given the complexity of the problem, and the need to involve the semi-independent MOHs, the system it created was, of necessity, cumbersome[66] but once it was in place, the GRO could abstract mortality data by sanitary district in the knowledge that their results would match those produced locally.[67]

Nor did these innovations exhaust the impact of the introduction of machine tabulation. The greater processing power of the GRO allowed it to introduce the new tables, and enhancements to old statistical series, which, as already discussed at the beginning of the last chapter, were such a feature of the *Annual reports* in the years immediately preceding the First World War.[68] Thus, the positive impact of eugenics on the GRO can be seen in terms of contributing to the unblocking of the barriers to technological innovation, rather than in simply raising the institution's public profile. The Treasury was more likely to be impressed by the introduction of machinery that it was told would cut unit costs, than in intellectual debates.

[65] PRO: RG 29/4, p. 176; PRO: MH 78/114 GRO. Co-ordination of work with Ministry of Health and appointment of Registrar General, 'Note by Dr Stevenson' dated 25th January 1920.
[66] For a description of the system, see: Higgs, 'The statistical Big Bang of 1911', pp. 424–5.
[67] PRO: RG 26/34 Transferable Deaths – policy (Dr Buchanan), 1921, and 1940; RG 26/35 Transferable Deaths – Correspondence, 1910–43, letter of 7 Nov 1911 from the Public Health Department of Notts County Council to Newsholme.
[68] GRO, *73rd ARRG for 1910* , pp. vii–viii.

The consequences of technological innovation – organisational change

The routinisation of the use of this technology also had a significant impact upon the structure of the GRO. Most of the staff permanently employed in the Statistical Department prior to 1911 were not trained statisticians. Their work was, in practice, a mixture of manual dexterity and rudimentary scientific knowledge – they ticked boxes according to predetermined classification systems. This process involved elements of analysis since the placing of the ticks involved a certain amount of discrimination, and a familiarity with census classifications and nosologies. The complex internal gradations within the Department reflected varying degrees of experience and skill in this work, although a considerable portion of the time of the superintendent of statistics, and of his higher officers, was spent in supervising these activities.

Machine tabulation radically disrupted this hierarchy, since the primary task of punching involved little, if any, discrimination. In the latter process a number was attached to a term, whilst in the manual system the name had to be positioned within the topography of a classification system. The difference is analogous to that between using a gazetteer to find in which country a place lies (adding one name to another), and actually being able to find that place on a map. The work of the new breed of female and juvenile 'punchers' had a minimal intellectual component requiring neither specialist training or critical supervision. As a result of these changes the Statistical Department was split in two in the immediate aftermath of the First World War. A new section bearing the old title was established, staffed by large numbers of machine operators and coders ('operational staff') on relatively low and undifferentiated wages. The higher officers of the new Statistical Department (the 'controlling staff') inherited the supervisory and managerial functions that had taken up so much of the time of Farr and his successors. T. H. C. Stevenson, the former Superintendent of Statistics, was removed to a new 'Professional Department', to be but one of a number of 'statistical officers' specialising in the

State medical statistics in the Edwardian era 177

analysis medical and census statistics.[69] By the 1950s the professional statistical work of the GRO in London had been geographically separated from the coding and punching, which was now concentrated in a common services department in Southport, Lancashire.[70] A greater gap had thus opened up between a scientific elite and a mechanical substrata. The replacement of unionised male clerks by, theoretically, more docile women and children was also a boon to managers in terms of Civil Service politics.[71]

Such a flattening of organisational hierarchies will be familiar to anyone who has studied the recent literature on the effects of the introduction of modern information technology into business organisations.[72] What is going on in modern IT-based organisations is somewhat different, however, to the changes in the GRO after 1911. In order to understand this difference one might draw upon a distinction made by Shoshana Zuboff between 'automating' and 'informating' processes. In the former machines replace skills, whilst modern information technology provides information for an intelligent workforce to use. Both strategies may lead to the removal of sections of the workforce but whilst automation leaves the skill levels of those left behind to tend the machines that much lower, this is not the case in informating.[73] In the GRO, the automation of processes led to the removal of semi-skilled workers and their replacement with the relatively unskilled. In modern information-based organisations, semi-skilled workers are also removed but leave a relatively highly skilled workforce behind.

In the long-run, the establishment of a new cadre of professional statisticians within the GRO, divorced from the day to day supervision of data-processing, may have helped to improve

---

[69] PRO: RG 20/6 Re-organisation of the Higher Establishment of the GRO, Min of Health and Treasury Correspondence, 1920–23; PRO: RG 26/50 Statistical Branch: notes on staffing, reorganisation, analysis of duties, the need for cultivation and preservation of expert knowledge and comparison with other Statistical Departments. 1937–42, 'Notes of Staffing of 'P' and 'S' Branches, p. 1.
[70] PRO: RG 20/50 Staff organisation chart January 1 1951.
[71] Zimmeck, 'Strategies and stratagems for the employment of women'.
[72] See, for example, Bikson, 'Organisational trends and electronic media: work in progress', pp. 57–9; Simpson, 'The management of electronic information resources'.
[73] Zuboff, *In the age of the smart machine*, pp. 9–11.

the scientific credentials of the GRO's senior staff. It is debatable, however, whether the dissociation of data processing and statistical analysis of itself improved the quality of either activity. Data processing became a purely mechanical job, the end-purpose of which was only dimly understood, and thus combined tedium with apparent futility. This may have removed the ability, or desire, to use initiative in bringing forward problems with the data. At the same time, there was a potential for the statisticians to be removed from intimate contact with their raw materials, and thus to lose a 'feel' for their limitations.[74]

The statistical limitations of the GRO in the early twentieth century

But whilst eugenics helped to revolutionise the data processing capacity of the GRO, it also, in the short term, undermined the professional standing of its statistical cadre. Whereas the impact of eugenics on the GRO has been discussed here mainly in terms of its implications for social policy, one should not overlook the manner in which it contributed to the development of new statistical methodologies. Eugenics in the hands of Francis Galton was intended to be a mathematical science, rather than just an ideology, and the theories of correlation, sampling, and statistical significance to which, in part, it gave rise were a profound advance on the use of ratios, rates and raw numbers favoured by older statistical institutions such as the GRO.

Galton and Pearson helped to institutionalise the new statistics, as well as providing its early intellectual break-throughs. Pearson had become the Goldsmid professor of applied mathematics and mechanics at University College, London, in 1884. Here he came into contact with W. F. R. Weldon, the College's professor of zoology from 1890, and together with him founded the 'Biometric School' to apply the new statistics to human and animal biology.[75] Pearson founded *Biometrika* in 1901, along

---

[74] This was certainly the belief of some statisticians working in government in the 1970s: Government Statisticians Collective, 'How official statistics are produced', pp. 142–6.

[75] Magnello, 'Karl Pearson's Gresham lectures'.

with Weldon and Galton, and under his editorship it became one of the major outlets for published work on statistical theory in Britain. In 1905 Francis Galton gave the University of London £1,500 to establish a Eugenics Record Office, and £500 annually until his death in 1911 for eugenic research. In 1906 Galton asked Pearson to take over the Eugenics Record Office, which was renamed the Laboratory of National Eugenics. The Eugenics Laboratory, together with a 'Biometric Laboratory', funded from Pearson's other resources, and oriented towards statistical theory, became the institutional base for the Biometric School as a combined Department of Applied Statistics. It should be noted that the Biometric Laboratory was Pearson's main concern, and it was here that he developed the technical aspects of his statistical work, whilst the Eugenics Laboratory was relatively peripheral to his activities.[76] When Galton died he left the residue of his estate to the University of London to establish a Galton professorship of eugenics, and recommended that Pearson should be appointed to the post.[77]

Pearson used the School to bring together young men and women who were interested in eugenics and/or statistical theory. George Udny Yule, for example, was appointed as Pearson's demonstrator in 1893, and went on to become an assistant professor of applied mathematics in 1896, and Newmarch lecturer in statistics at University College in 1902. From 1912 he was university lecturer in statistics at Cambridge.[78] Similarly, Major Greenwood was supposedly so fired by reading Pearson's *Grammar of Science* whilst still a medical student at the London Hospital, that he decided to devote himself to biometrics. He made contact with Pearson and studied under him at University College during the academic year 1904/5. Greenwood then went on to take up a research scholarship from the British Medical Association in 1905, and became the demonstrator in the physiological laboratory of Leonard Hill at the London Hospital Medical School. In 1910 he was appointed statistician at the Lister Institute of

---

[76] Magnello, 'The non-correlation of biometrics and eugenics'.
[77] MacKenzie, *Statistics in Britain*, pp. 101–4.
[78] *Concise Dictionary of National Biography*, p. 3330.

Preventative Medicine, and began a distinguished career in medical statistics, particularly in epidemiology and public health.[79] During the First World War Greenwood worked for the Welfare and Health Section of the Ministry of Munitions on the incidence of phthisis among munitions workers, and on the wastage of labour in munitions factories employing women.[80] On the establishment of the Ministry of Health in 1919 he was seconded to it as a statistical officer.[81]

Although influenced by Pearson's statistics, both men quickly became sceptical of his eugenic ideas.[82] Indeed, Yule worked on infant mortality with both Arthur Newsholme and George Newman, the chief medical officer at the Board of Education, in the years before the First World War.[83] As already noted, Pearson appears to have maintained a distinction between his statistical work in the Biometric Laboratory and his eugenics work in the Eugenics Laboratory[84], and in time the eugenic and statistical components of the Biometric School formally parted company. In the 1930s, left-wing intellectuals such as J. B. S. Haldane and Lancelot Hogben used mathematical techniques to undermine the scientific pretensions of eugenics, and laid the groundwork for the modern study of genetics. Haldane held a separate chair of biometry in London University from 1935, and in 1943 succeeded R. A. Fisher to Pearson's old position as the Galton professor of eugenics. In 1945 Lionel Penrose, a colleague of Haldane and Hogben, and a non-eugenicist, took over the Galton chair. In 1957, when Haldane left the University, the joint Department of Eugenics and Biometry passed into Penrose's hands. During Penrose's tenure no eugenics were taught in the department; the name of the house jour-

---

[79] Hogben, 'Major Greenwood'; Matthews, *Quantification and the quest for medical certainty*, p. 104; MacKenzie, *Statistics in Britain,* pp. 110–11.

[80] PRO: FD 2/4 Fourth Annual Report of the Medical Research Committee, 1917–1918 (Dated 18 October 1918), pp. 19–20, 66.

[81] For Greenwood's career see, Edward Higgs, 'Medical statistics, patronage and the state'.

[82] Matthews, *Quantification and the quest for medical certainty* , pp. 117–8; MacKenzie, *Statistics in Britain,* pp. 106, 111, 173–4.

[83] Soloway, *Demography and degeneration,* pp. 150–1.

[84] Magnello, 'The non-correlation of biometrics and eugenics'.

nal was changed from *Annals of Eugenics* to *Annals of Human Genetics*; and in 1963 the title of the chair was changed to the Galton professorship of human genetics. The new sciences of statistics and genetics had finally cast off the influence of their eugenic inheritance.[85]

Roger Davidson has posited a reluctance amongst Civil Service mandarins to introduce these new statistical methods into Whitehall. They were seen as too advanced for ministers and the public to understand, and were opposed by established statistical cadres, afraid of the undermining of their expertise.[86] The senior officers of the GRO were, however, aware of the extent to which the Office had failed to keep abreast of the new statistical techniques. Whereas Farr had been president of the Royal Statistical Society, one finds Mallet confiding to the Treasury in December 1911 that, 'The Statistical Branch as at present staffed is in my opinion hardly strong enough in its upper ranks for the work which it will be required to do under its present progressive chief.' He also sought permission to employ outside experts to give 'advice on matters connected with recent developments in the mathematical treatment of statistics . . .'.[87] The following month Mallet was asking for authority to pay 'Edny Ule' and E. C. Snow £50 for such advice.[88] In March 1912 both Yule and Edwin Cannon attended a conference at the GRO, 'on the reform of the population tables'.[89] The failure to introduce the statistical innovations of the Biometric School into the GRO at this date may not, therefore, have reflected outright opposition to change. Rather, the lack of suitably qualified recruits may have been the crucial factor.[90] As Major Greenwood claimed, somewhat vaingloriously, in the *Lancet* in 1913, 'The number of biometrically trained medical men is still small, and, as far as I am aware, I am

---

[85] Mazumdar, *Eugenics, human genetics and human failings*, pp. 146–95, 253, 318 n. 169.
[86] Davidson, 'Social intelligence and the origins of the welfare state', p. 47.
[87] PRO: RG 29/4, pp. 191–2.
[88] Ibid., p. 199; RG 29/8, p. 71.
[89] PRO: RG 29/4, p. 212.
[90] Davidson, *Whitehall and the labour problem*, pp. 235–6.

the only one in this country holding a post expressly created to further the application of biometric methods in medicine and pathology'.[91]

Similarly, whereas Farr had been an acknowledged expert in actuarial matters, the twentieth-century GRO increasingly looked to outside experts to construct its life tables. Thus, those that made up the first volume of the 1911 *Decennial supplement*, published in 1914, were produced by George King, the vice-president of the Institute of Actuaries.[92] In the introduction to this volume, Mallet noted that life tables had hitherto been produced by his own staff but that since the construction of life tables 'involves work of a highly technical nature ... it appeared desirable to obtain the advice of an expert skilled in such work'.[93] King noted in the body of his report that Farr's life tables, and the GRO's Life Tables 5 and 6 gave an underestimate of mortality at advanced ages.[94] King also consulted Yule on the finer points of his calculations for the tables.[95]

Much the same could be said of the second volume of the 1911 *Decennial supplement*, which was published after the First World War. This was a set of abridged tables that gave certain functions at a few selected points of age instead of at every year of age as in an extended life table.[96] According to Mallet, in its attempts to establish methods for creating better abridged tables the GRO, 'became aware that Dr E. C. Snow, with whom, as with other statisticians, frequent consultations had taken place over matters of statistical interest, was engaged in private investigations having the same object in view.' The Office had made little progress in the matter, so it placed its own calculations at Snow's disposal.[97] All subsequent life tables produced by the

---

[91] Quoted in Mathews, *Quantification and the quest for medical certainty*, pp. 111–12.
[92] GRO, *Supplement to the 75th annual report of the Registrar General:. Part I: Life Tables.*
[93] Ibid., p. iv.
[94] Ibid., p. 2.
[95] PRO: RG 26/77 Life tables: correspondence (1912–1915) G King, Actuary, letter to Stevenson of 28 November 1912.
[96] GRO, *Supplement to the 75th annual report of the Registrar General:. Part II: Abridged Life Tables.*
[97] Ibid., pp. iii–iv.

GRO in the inter-war period were to be the work of the Government Actuary, Sir Alfred Watson.[98]

The position of the GRO as the central body in England and Wales for the statistical study of mortality was also being undermined in this period by the creation of other state institutions dedicated to similar activities. The establishment of the National Insurance scheme in 1911 increased the need on the part of central government for actuarial advice to deal with the assessment of the soundness of the participating societies. An expansion of the actuarial activities of the GRO to meet this requirement was not even considered. Instead, the National Health Insurance Joint Committee turned in 1912 to a committee made up of members of professional actuarial bodies: Watson, the Committee's chief actuary; G. F. Hardy, ex-president of the Institute of Actuaries; G. J. Lidstone, vice-president of the Institute of Actuaries; Gordon Douglas, president of the Faculty of Actuaries in Scotland; and D. C. Fraser, actuary of the Royal Insurance Co.[99] This committee was, in time, to grow into the Government Actuaries Department, with Watson at its head.

The National Insurance Act also provided for a levy on contributions to undertake medical research.[100] This led to the establishment of the Medical Research Committee, which became the Medical Research Council (MRC) in 1920.[101] The MRC was not subordinated to a government department, as in the case of the GRO, and was given a block grant from the Treasury over which it had control.[102] Under Sir Walter Fletcher, its first secretary, its orientation was firmly towards the academic pursuit of knowledge, rather than the servicing of

---

[98] GRO, *The Registrar General's decennial supplement, England and Wales 1921. Part I. Life* Tables; GRO, The *Registrar General's decennial supplement, England and Wales 1931. Part I. Life Tables*.

[99] PRO: ACT 1/1 Actuarial Advisory Committee on the operation of the National Insurance Act, 1911: Appointment and reports: minute of appointment, 29 January 1912.

[100] Thomson, *Half a century of medical research. Volume 1*, pp. 11–21.

[101] For the origins and early development of the MRC see, Higgs, 'Medical statistics, patronage and the state', passim.

[102] Ibid., pp. 22–65; Austoker, 'Walter Morley Fletcher and the origins of basic biomedical research policy'.

government policy making. To this end it undertook its own medical research but also funded similar activities in universities and hospitals.[103] As already noted, when the MRC established a National Institute for Medical Research in 1914, it contained a Statistical Department under John Brownlee, another pupil of Pearson.[104] Brownlee's department was intended to:

> consist of persons in the permanent employment of the scheme who will be engaged in enquiries relating to diet, occupation, habits of life and other matters bearing upon the incidence of disease, and who will collect and deal with all types of vital statistics including the distribution of disease, the relative frequency of special types of lesions in diseases such as tuberculosis, and in general with all statistical investigations useful either as preliminary to research or confirmatory of its results.[105]

It is also possible that the MRC Statistical Department was envisaged as the means of undertaking inquiries into causes of excessive sickness under section 63 of the National Insurance Act. This allowed for the holding of enquiries to determine if excessive claims on the scheme were due to the living conditions or nature of employment of insured persons. In practice the arrangements under this clause proved uninforceable[106] but the National Insurance Joint Committee certainly saw the need for statistical enquiries of this sort.[107] The first *Annual report* of the MRC noted that its Statistical Department would 'probably have to consider and advise how the statistical material provided for under the Insurance Act should be dealt with.'[108] The very

---

[103] For the MRC's research programme, see: Thomson, *Half a century of medical research. Volume 2.*
[104] Thomson, *Half a century of medical research, Volume 1,* pp. 9–29, 110.
[105] Ibid., p. 29.
[106] For the weaknesses of the Excessive Sickness Clause, see: Eyler, *Sir Arthur Newsholme,* p. 231.
[107] PRO: ACT 1/1 Actuarial Advisory Committee on the operation of the National Insurance Act, 1911: appointment and reports, 1912–13; memorandum on 'Statistics and returns'.
[108] PRO: FD 2/1 First Annual Report of the Medical Research Committee, 1914–1915 (Dated 18 October 1915), p. 6.

first special report published by the MRC was indeed one by Brownlee on the excessive incidence of phthisis in the boot and shoe industry.[109]

Given the rather ineffectual leadership of John Brownlee the MRC's Statistical Department was hardly a threat to the GRO.[110] In the aftermath of the First World War, however, the MRC became the focus of advanced statistical research in medical matters within government. Despite the innovations in classification systems associated with T. H. C. Stevenson, and in data management technologies, in the early twentieth century, the GRO was ceasing to be a centre of statistical excellence. The relative weakness of the Office in such methodologies was to help to undermine the independence of the GRO when it passed under the Ministry of Health in 1919. These developments will be the subject of the next chapter.

---

[109] PRO: FD 4/1 The first report of the special investigation committee upon the incidence of phthisis in relation to occupation. The boot and shoe industry.

[110] For the ineffectual nature of Brownlee, see: Higgs, 'Medical statistics, patronage and the state', pp. 327–8.

7

1914–1951: registration and statistics in total war and total welfare

Citizenship and World War 1 – the limits of the registration system

The First World War was, at one level, a serious blow to the GRO. The enlistment of clerks in the armed forces, and the imposition of new duties on those who remained, significantly reduced the Office's statistical capabilities.[1] By 1917 most of the male clerks of serviceable age had left for the Front, or been loaned to other departments for war work. Their place was taken by untrained women but the pressure of work falling on the senior staff was still considerable.[2] The *Annual reports* for the war years were delayed as a consequence, and had shorter texts and fewer tables than their pre-war predecessors (see Figure 3:2). At the same time, most of the results of the 1911 fertility survey, and of the volumes of the *Decennial supplement* for 1911, would not be published until the Armistice had been signed, or even until the 1920s.[3]

By contrast, the importance of the role of the GRO, and of the registration system in general, in underpinning rights and enforcing obligations reached new heights. As already noted, the Office had to trace marriage certificates for those claiming separation allowances and widows pension, and it also assumed

---

[1] GRO, *76th ARRG for 1913*, p. viii.
[2] GRO, *78th ARRG for 1915*, p. viii.
[3] *Census 1911, Vol. XIII, Fertility of marriage report, Part 1*; *Census 1911, Vol. XIII, Fertility of marriage report, Part 2*; GRO, *Supplement to the 75th annual report of the RG: Part II: Abridged Life Tables*; GRO, *Supplement to the 75th annual report of the RG: Part III: Registration Summary Tables*; GRO, *Supplement to the 75th annual report of the RG: Part IV: Mortality of Men in Certain Occupations in the Three Years 1910, 1911, and 1912*.

responsibility for the Central Register of War Refugees.[4] But it was, above all, its role in the national registration system that taxed the GRO to its limits. By 1915 it was plain that the war with Germany could only be won if the entire human resources of the country could be mobilised.[5] A series of reports produced by the Jackson and Landsdowne Committees on National Registration called for the establishment of a system that would facilitate military conscription but also enable the claims of industry and agriculture to be taken into account.[6] Under the resulting 1915 National Registration Act, the Registrar General became the central registration authority, with the metropolitan and municipal boroughs, and urban and rural district councils, acting as the local registration authorities. By January 1917 the GRO was employing 59 members of its staff on national registration work.[7] In the later stages of the War, the GRO was also responsible for issuing ration books and sugar tickets to the public via the local registrars.[8] The overall success of these administrative arrangements encouraged T. H. C. Stevenson in September 1916 to suggest the possibility of continuing the national registration system after the War.[9] As a result of Stevenson's musings, and with the active support of Sir Bernard Mallet, a Committee on National Registration was set up in May 1917 by the President of the

[4] GRO, *77th ARRG for 1914*, p. viii.
[5] Guinn, *British strategy and politics*, pp. 85, 179–81; Winter, *The Great War and the British People*, pp. 25–48; French, *British strategy and war aims*, pp. 116–31; Grieves, *The politics of manpower*, pp. 19–24.
[6] PRO: RG 28/8 1915 National Registration (Jackson) Committee: minutes; first and second interim reports: first interim report (6 August, 1915), p. 2; RG 28/9 1915 National Register (Landsdowne) Committee: minutes, reports and papers: interim report (3 September, 1915), p. 2.
[7] PRO: RG 28/1 Preparations for, and administration of, the National Registration Act 1915–18 Vol. I, p. 24.
[8] PRO: RG 48/585 1917–1918 Duties in respect of rationing schemes: correspondence with Ministry of Food and Registrar General for Scotland.
[9] PRO: RG 28/3 National Register 1915–19. Vol. III. Committee on National Registration 1917–18: Prt I Correspondence, memorandum of September 1916. Simon Szreter implies that the idea came from the Registrar General Sir Bernard Mallet: Szreter, *Fertility, class and gender in Britain*, p. 267, n. 116. This appears to be based on a claim made by Mallet in the 1920s but the idea appears actually to have originated with his subordinate. See also, Mallet, 'The organisation of registration', pp. 22–3.

LGB. The Committee recommended the establishment of a permanent register but the proposal foundered because of the cost and fears over civil liberties.[10]

This should not be taken to imply, however, that the registration functions of the GRO ceased to employ the majority of its officers. As can be seen from Table 7:1, the preponderance of staff in the Registration Branch of the Office was maintained in the inter-war period. However, the failure of the GRO to carve out a distinctive role for itself as *the* centre for the recording of citizens' rights and national manpower reveals the limits to the centralising tendencies of the British State in the aftermath of the Great War. This failure left the GRO vulnerable to assimilation by another body that was promoting integration within Whitehall in this period, the newly constituted Ministry of Health.

Table 7:1. Staff of the GRO as of 1 April 1939

Registration Branch		Statistical Branch	
Clerical grades	181	Clerical grades	81★
Other grades	139	Other grades	1
		Census Branch	
		Clerical grades	24
		★ Includes punch operators as 'clerical assistants'	

Source: PRO: RG 20/13 Staff statistics and proposed cuts in staff 1939–1943

The Ministry of Health and the fall of Sir Bernard Mallet

In the Edwardian period demands had been building for greater leadership from the centre in matters of public health. Bellamy links this, in part, to the development of larger county councils

---

[10] For the history and context of these proposals see, Higgs, *The Information State in England*, pp. 134–40.

that had taken political decisions out of the hands of small, local property holders of limited vision and expertise. Political lobbies had also developed to press for slum clearance and poor law reform, and were increasingly stressing the need for compulsion from the centre.[11] There was also a desire to reintegrate health care provision that had become fragmented by the establishment of the national insurance system and new child health services outside the LGB. The former was the responsibility of the National Health Insurance Commission (NHIC), whilst the latter reflected the expanding reach of the Board of Education, and of its chief medical officer, George Newman. This reintegration could not be achieved via the LGB since it was the hostility of politicians and reforming civil servants to the perceived incapacity of the Board that had led to the fragmentation in the first place.[12]

The unification of the diverse bodies responsible for the health of the nation was achieved in the aftermath of the First World War by the passage of the 1919 Ministry of Health Act. As the first *Annual report of the chief medical officer of the Ministry of Health* put it, the Ministry was, 'a new central authority created for the purpose of supervising the health of the people *as a whole* and for unifying and simplifying the central agencies working on its behalf.'[13] The Act led to the amalgamation of the NHIC and the LGB, but hardly on the latter's terms. Sir Robert Morant, the chairman of the NHIC became permanent secretary of the new ministry, whilst the permanent secretary of the LGB, Sir Horace Monro, was forced to resign. Similarly, the post of chief medical officer went to George Newman rather than to Sir Arthur Newsholme, who also left the Civil Service.[14] The new Minister of Health was Dr Christopher Addison, who had helped to steer the National Insurance Act through Parliament,

---

[11] Bellamy, *Administering central-local relations*, pp. 237–51.
[12] Ibid., p. 252; Honigsbaum, *The struggle for the Ministry of Health*, pp. 20–3; Eyler, *Sir Arthur Newsholme*, pp. 316–33.
[13] Ministry of Health, *Annual report of the chief medical officer of the Ministry of Health 1919–20*, p. 8.
[14] Honigsbaum, *The struggle for the Ministry of Health*, p. 55; Eyler, *Sir Arthur Newsholme*, pp. 316–37.

and was a strong supporter of Newman.[15] The GRO joined the rest of the LGB in the new Ministry of Health, and was also infiltrated by ex-NHIC staff in the form of the appointment of Sylvanus Vivian, an assistant secretary at the NHIC, to the post of 'deputy' Registrar General in November 1919. Vivian was so closely identified with Morant that he was known in Whitehall as 'Morant's little Sylvanus'.[16]

The Ministry of Health's view on the constitutional and statistical position of the GRO was that its independence had to be curtailed. An undated Ministry of Health memorandum, probably of 1919, entitled 'The relations, past and future, between the Registrar General and the Minister of Health', gave a history of the Registrar General's office that could stand as a useful summary of some of the detailed arguments of the present work. It noted that:

> The office of Registrar General was created in times in which the modern doctrine of ministerial control over, and responsibility for, the executive was by no means established. In the first instance the Registrar General was an autonomous potentate, unrelated to any minister of the Crown save by the duty imposed on him to report to the secretary of state (afterwards the LGB). As the doctrine of ministerial control gathered force, it became increasingly usual for the statutes affecting the functions of the Registrar General to prescribe that this or that specific function should be exercised by him subject to the approval of the LGB; and at the present day the functions of the Registrar General consist of an aggregate of specific statutory functions in which his authority is subject to the approval of the Minister of Health, and a residuum of indefinite functions in which he is autonomous....

The memo went on to suggest that it was now possible for the Minister of Health to assume towards the Registrar General the

---

[15] Turner, "Experts' and interests', p. 210; Eyler, *Sir Arthur Newsholme*, pp. 334–5.
[16] Stacey, 'The Ministry of Health', p. 43. This thesis gives a detailed account of the Ministry's take-over of the GRO: Stacey, 'The Ministry of Health', pp. 89–99.

normal relation of a minister to the head of a department responsible to him.[17] In August 1919 Lloyd George agreed that Addison should be given complete responsibility for the GRO, and a free hand over all appointments to it.[18]

The Ministry's line of reasoning was much the same with regard to statistical production. During the war a sub-committee of the Reconstruction Committee established to consider the establishment of a Ministry of Health had already concluded that the creation of a Ministry would be a means of ensuring that the GRO's reports would be prepared on a common basis with the those of other departments.[19] The report of the Machinery of Government (Haldane) Committee of 1918 had also stressed the need for research and information to be available to central policy makers. Morant, who had begun his career in a specialist intelligence division of the Board of Education, and had served on the Haldane Committee, was passionately interested in research and intelligence, and committed to its policy uses.[20] He was writing to Mallet as early as June 1919 indicating that the Ministry was approaching the Treasury with an application for more resources for the GRO, 'by reason of important coming developments in the sphere of registration of many different kinds, which would need to be continually linked up with health insurance, with other functions of the Ministry of Health, and possibly with other registration necessities such as unemployment insurance and so forth.'[21] The need to centralise and co-ordinate information, to create what A. L. Bowley called a 'central thinking office', was a general feature of Western states at this date.[22]

---

[17] PRO: MH 78/114 GRO co-ordination of work with the Ministry of Health and appointment of Registrar General: 'The relations, past and future, between the Registrar General and the Minister of Health', pp. 1–3.
[18] PRO: T 162/82 General Register Office: Registrar General, Accounting Officer, and Statisticians Appointment: letter of 2 December 1920 from Robinson to Warren Fisher.
[19] PRO: FD 5/1 Proposed Ministry of Public Health and Local Government: correspondence with MRC, 1916–17: Report of the sub-committee on a Ministry of Health, pp. 24–5.
[20] Stacey, 'The Ministry of Health', pp. 49–50, 86.
[21] PRO: MH 78/114, letter of 23 June 1919.
[22] Beaud and Prévost, 'La forme est le fond'.

On 3 November 1919 the Ministry wrote to the Treasury noting that the Minister of Health proposed to take the earliest opportunity of overhauling the organisation and staffing of the GRO.[23] Two days later Addison appointed Vivian as deputy Registrar General with the remit of examining the organisation of the Office and reporting back to the Ministry's director of establishments. The Registrar General was merely to be 'informed of all proposals and developments.'[24] In January 1920, a meeting between Addison, Morant and Vivian decided that 'pending the transfer to the Ministry of Health of full control, de jure as well as de facto, over the Registrar General's department, all lines of communication should pass through the deputy Registrar General . . .'.[25] A letter from the Ministry's director of establishments to the Treasury in November 1919 was already hinting that Vivian's position as deputy Registrar General 'would be subject to reconsideration on the occurrence of a vacancy in the post of Registrar General'.[26] The staff side of the GRO's Whitely Council, one of the bodies set up by Lloyd George in each department to facilitate liaison between management and trade unions, saw this as the beginning of 'the submergence of this Office as an independent unit', and wrote to Vivian deploring the effects on promotion.[27]

Sir Bernard Mallet was deeply unhappy at these developments, and complained to the Treasury as early as August 1919 that Vivian's activities were over-riding the provisions of the 1836 Registration Act, and reducing the GRO to 'a mere branch of the Ministry of Health'.[28] Mallet was plainly in an untenable position and vacillated between bluster and acquiescence. As Morant minuted the Minister in February 1920:

[23] PRO: MH 78/114, letter of 3 November 1919.
[24] Ibid., memo of 7 November 1919.
[25] Ibid., 'Note of Dr Addison's. Discussion of Registrar General's department'.
[26] PRO: T 1/12404/46946, letter from Woodgate to Treasury of 3 November 1919.
[27] PRO: RG 20/77 1919–1920 Whitely Council: representations from the Staff Side regarding reorganisation and promotion prospects: letter to Vivian of 14 November 1919.
[28] PRO: T 1/12404/46946, letter of Mallet to R S Meiklejohn, 10 August 1919.

Sir Bernard half-heartedly puts a point or two weakly, and then, after discussion, seems to indicate assent or withdraws his dissent. And then, weeks later, produces some long rigmarole of objections, with an insinuation that he has been hardly treated in not being allowed to be heard in matters within his sphere before decisions are reached. I cannot too strongly say that this is precisely the way he behaved all through last year towards myself and Sir John Anderson when the original questions of having Mr Vivian as deputy Registrar General was first mooted.

Mallet was accused of 'doing all he could to poison the minds of the Treasury officials'.[29] The following month he sent an ultimatum to the effect that unless Vivian undertook to be wholly subordinate to him he would resign.[30] By the end of the year Mallet was gone, and Vivian became Registrar General in name, as well as in practice, on 1 January 1921. However, Vivian still retained his position as principal assistant secretary in the Ministry of Health[31], and was to remain in post until 1945.

Vivian's eventual recommendations with regard to the GRO's statistical activities struck a blow at the very heart of the Office's intellectual independence. The role of the Statistical Department was henceforth to be ancillary to the activities of the rest of the Ministry of Health:

> The Registrar General's department must not be the department <u>responsible</u> for initiating or originating research upon its statistics; for if it did so on its own initiative it might be duplicating, or cutting across the work of other departments properly charged with such duties....

The 'expert' interpretation of the GRO's data was to be done by others – medical data in the Ministry of Health; labour statistics in the Labour Department of the Board of Trade; and actuarial

---

[29] PRO: MH 78/114: Morant to Addison, 5 February 1920.
[30] Ibid.: letter of Vivian to Addison, 29 March 1920.
[31] HMSO, *Imperial calendar for 1925*, p. 53.

data by trained actuaries. The GRO could draw attention to noteworthy trends in its statistical series but was to steer clear of drawing inferences from the data that had a bearing on policy and might cause trouble for the Ministry.[32] Vivian showed his desire to remove the overlap between the Ministry's statistical work and that of the GRO by proposing the integration of the latter's weekly returns and the Ministry's internal departmental listing of infectious diseases.[33] In order that the GRO might be suitably advised on the scope and presentation of its output by its 'clients' – other parts of the Ministry and other government departments and public organisations which utilised its products – Vivian suggested that a standing statistical council should be established within the Ministry. The model of advisory councils or committees was one which Addison and Lloyd George introduced widely in the administrative reforms.[34] This was the culmination of the long-term reorientation of the GRO from local propaganda to the provision of data for central policy making. The contemporary relaunching of the GRO's main publication as the anonymous *Statistical review*, which ceased to be a Parliamentary Paper, fits into this inward-looking strategy.

There was, moreover, a general feeling within the Ministry that the GRO's Statistical Department was not qualified to undertake the tasks to be placed upon it. In January 1920, Vivian declared to Morant that in his opinion, 'the fact has clearly emerged that the GRO has never been equipped, and was never intended to be equipped, with an establishment suitable for dealing with the more responsible aspects of its work.'[35] In June 1920 the Ministry was complaining to the Treasury that delays in the production of statistical reports made it obvious, 'that the present professional staff is not adequate to perform even the

---

[32] PRO: MH 78/114: minute from Vivian to Newman [June/July 1921].
[33] PRO: RG 26/37 Amalgamation of weekly infectious diseases and the Registrar General's weekly returns 1921–23.
[34] PRO: MH 78/114: note on a conversation between Newman and Vivian, 21 September 1920; Turner, "Experts' and interests'.
[35] PRO: RG 20/80 Organisation and staffing 1920; memo from Vivian to Morant, 19 January 1920.

normal statistical work which devolves upon it ...'.[36] The question was, however, whether it was better to remedy the situation by absorbing the professional staff of the GRO into a 'medico-statistical department' in the Ministry of Health, or by beefing up the GRO through the introduction of an officer trained in the new statistical techniques.

Major Greenwood in the Ministry favoured the former course, and the idea was still being considered in 1923, when T. H. C. Stevenson was said not to be opposed in principle to moving to Health.[37] It was the option to reform the GRO, however, which was most rigorously pursued. Vivian's preferred strategy was to import, 'some picked young men direct from (say) Cambridge, who had shown a leaning towards the statistical side of mathematics, and whose introduction would help to train up capable statisticians for the public service and to satisfy a notorious deficiency in our present-day scientific equipment.'[38] In the summer of 1920, therefore, Vivian offered the post of 'super-statistician' in the GRO to Udny Yule, who turned it down.[39] Newman and Vivian then decided that Major Greenwood should be transferred to the GRO as a, 'statistical "superman" to undertake the professional supervision of the professional statistical staff of the Department'.[40] The prospect horrified Greenwood, and he complained to his friend Sir Walter Fletcher that as soon as 'I pass under Vivian my scientific usefulness ceases'. Together they began to hatch a scheme to remove Greenwood from the reach of the Ministry by seconding him to the MRC's National Institute for Medical

[36] PRO: T 162/1 Health Ministry: Registrar General's department reorganisation (higher staff), letter of 7 June 1920.
[37] PRO: RG 20/83; Creation of a Department of Vital and Medical Statistics; memo by Major Greenwood, November 1919; PRO: FD 5/91 Professor Major Greenwood: appointment, funding for research projects; letter of Greenwood to Fletcher 23 March 1920; PRO: RG 20/38 Relative functions of the GRO superintendent of statistics and the chief medical officer (Ministry of Health); letter of Buchanan to Newman, 26 December 1923.
[38] PRO: MH 78/114 'Registrar General's department. Memorandum on reorganisation of duties'.
[39] PRO: FD 5/91; letter from Major Greenwood to Sir Walter Fletcher of 17 June 1920.
[40] PRO: RG 50/4 1920–21 Correspondence and papers: proposed consultative committee on statistics; Electoral Division of the Ministry of Health.

Research.[41] Eventually V. A. P. Derrick from the Government Actuary's Department was appointed to take up one of the newly created posts of 'statistical officer' in the GRO's Professional Department alongside Stevenson.[42] This new arrangement had originally been suggested because it was considered impossible to subordinate Major Greenwood to Stevenson.[43] Above the professional officers, as head of the department, was placed an administrative assistant registrar.[44]

In practice, however, the GRO preserved a separate identity within the Ministry. This was partly because as soon as he became Registrar General, Vivian 'went native' and moved to build up his own institutional standing. Thus, in March 1921 he wrote to Sir Arthur Robinson, who became the Ministry's permanent secretary on Morant's death in 1920, noting that the Ministry had given the Treasury the impression that the separate existence of his office was to be extinguished, and its functions merged without distinction in the statutory functions of the Minister himself. Vivian countered that in fact all that was intended was the 'vesting in the Minister by statute the power of appointment and dismissal of the Registrar General and the control of the staff, in substitution for the present provisions of the law relating to the patronage of the office and to the Treasury control of the staff.' Robinson agreed that all that was required was a 'general tightening' of the relationship between the Ministry and the GRO, and the integration of common services.[45] The reforming drive of the Ministry had been undermined by the passing of Morant; the removal of Addison from the post of Minister of Health in 1921; and the general

---

[41] PRO: FD 5/91 letter from Major Greenwood to Sir Walter Fletcher of 17 June 1920; memo from Buchanan to Newman of 19 June 1920; unsigned letter [probably from Fletcher] to Newman, 29 June 1920; Higgs, 'Medical statistics, patronage and the state', pp. 332–3.

[42] PRO: T 162/1 Health Ministry: Registrar General's department reorganisation (higher staff).

[43] PRO: RG 50/4 1920–21 Correspondence and papers: proposed consultative committee on statistics; Electroral Division of the MH; undated memorandum.

[44] HMSO, *Imperial Calendar 1925–1934* (London, 1925–34); PRO: RG 20/84 Appointment of officers, 1932–35: letter from Vivian to Leggett of 8 June 1932.

[45] PRO: RG 20/82 Reorganisation: post of assistant Registrar General. Treasury official as accounting officer 1921–47; letter of Vivian to Robinson 15 March 1921; letter of Robinson to Vivian 6 June 1921.

retrenchment in government spending in the 1920s. The GRO had weathered the storm as a separate entity but now accepted its subordinate position. Whereas George Graham's assistance to other departments had been grudging and to some extent voluntary, one finds by the end of the 1930s that the GRO was seeking staffing parity with the statistical and professional branches of the Ministry of Labour on the grounds that 'it is our business to serve other departments' both within the Ministry of Health and across Whitehall.[46]

Similarly, the proposed internal statistical council to direct the work of the GRO failed to materialise, and in practice Stevenson had considerable control over the direction taken by the Office's published statistical output. In 1930 Vivian attempted to reassert the original concept of the relationship between the GRO and the Ministry's medical officers by asking the latter to draw up a programme of special investigations each Spring for inclusion in the forthcoming *Statistical review*.[47] Stevenson, moreover, does not appear to have ever accepted the post-1919 dispensation. Early in 1920, in the thick of the struggle between Mallet and Vivian, he had written an official minute justifying the control of the Statistical Department, 'by an executive superintendent holding a medical degree', rather than by a mathematician.[48] Subsequent events swept his post away but Stevenson continued to refer to himself in publications as 'Formerly Superintendent of Statistics, GRO'.[49] He was also discovered to have provided material for reviews in the *British Medical Journal* and the *Lancet* that criticised the new form of the *Statistical review* for 1921, and refused to divulge the names of the reviewers when pressed by Vivian. The new Registrar General warned Stevenson in no uncertain terms that, 'in the event of trouble arising out of these reviews his position might be difficult . . .'.[50] It may have been,

---

[46] PRO: RG 26/50 GRO statistical staff 1938; memorandum to the Registrar General 22 October 1937.
[47] PRO: MH 78/114; letter from Vivian to Robinson 5 February 1930.
[48] Ibid.; Note by Dr Stevenson, 25 January 1920.
[49] Stevenson, 'The social distribution of mortality from different causes in England and Wales', p. 382.
[50] PRO: RG 50/6 1923 G. S. P. Vivian's Correspondence: criticism of Annual Statistical Review 1921.

of course, that Stevenson's close association with Sir Arthur Newsholme, both as his assistant as MOH in Brighton and when the latter was at the LGB, had already made him a marked man under the new regime.

Stevenson was subsequently kept on a tight rein. When in 1923, for example, he was invited to a meeting of statisticians at the Health Section of the League of Nations in Geneva, Stevenson was sent a formal Ministry memo respecting his conduct there. He was instructed that since there were Bills before Parliament touching the form of the death certificate and the registration system as a whole, he 'must be careful to avoid any expressions of opinion which could be interpreted as implying any definite official attitude on the subject of the administrative system of death certification and the various suggestions for its amendment'. If a report was produced mentioning the medical aspects of registration, Ministry of Health officials wanted to see a copy of it.[51] This was a far cry indeed from William Farr's activities at the nineteenth-century international statistical congresses that had involved him drawing up vast plans for national and international data gathering. By 1930, Sir Arthur Robinson regarded Stevenson as a 'thoroughly discontented officer', and considered removing him from the GRO altogether.[52] The following year Stevenson died, the last independent head of the GRO's statistical branch in a line stretching back nearly a century to the appointment of Farr in the 1830s.

The nature of the GRO's statistical output in the inter-war period can certainly be seen in terms of a re-orientation to servicing Whitehall. The GRO's general lack of interest in following up the effects of social class on mortality, for example, was typical of the Ministry of Health as a whole. In the inter-war period the annual reports of the Ministry's chief medical officer did not address the subject, and the possible adverse effects of unemployment on mortality rates were

---

[51] RG 20/38 Relative functions of the GRO superintendent of statistics and the chief medical officer (Ministry of Health) 1920–1924.
[52] PRO: MH 78/114, letter of Robinson to Vivian 16 January 1930.

discounted.[53] The surviving correspondence between the GRO and Ministry on the latter subject reveals an eagerness on the part of the Registrar General to forward the generally positive line of the Ministry by providing it with statistical data to refute what were regarded as alarmist newspaper articles.[54] The emphasis of the *Statistical review* on overcrowding and population density, rather than social class, as explanatory variables also fitted in with Health's inter-war housing and slum clearance policy.[55] The Ministry, as the LGB before it, was more concerned to work within the administrative structures laid down in existing Acts of Parliament, and via liaison with local authorities, than to question the inequalities in the British social structure. The GRO now appeared on the national stage as a statistical spear-carrier for the officials of its parent department. It is here that Roger Davidson's arguments regarding the stultifying effect of central policy-making on statistical analysis have a degree of purchase.[56]

The work undertaken by the GRO in this period also showed a growing emphasis on national trends as opposed to local conditions. Rather than Farr's strictures on the insanitary activities of the burghers of Newcastle-upon-Tyne, the GRO in the early twentieth century concentrated on the general causes of infant and maternal mortality, and the effects of overcrowding and, occasionally, social class on the health of the nation. The central spatial variables used to investigate mortality in this period were not specific place or locality but settlement type

---

[53] Ministry of Health, *On the state of the Public Health. Annual report of the chief medical officer of the Ministry of Health for the year 1932*, pp. 16–43; Ministry of Health, *On the state of the Public Health. Annual report of the chief medical officer of the Ministry of Health for the year 1933*, pp. 206–21; Ministry of Health, *On the state of the Public Health. Annual report of the chief medical officer of the Ministry of Health for the year 1934*, pp. 15–17.

[54] PRO: RG 26/28 Mortality by place of occurrence – depressed areas. *Special Areas (Development and Improvement) Act 1934*; Dr Percy Stocks' report on 'Tuberculosis Mortality in the Urban and Rural Districts of Glamorgan and Monmouth' and 'Mortality in the Special Areas specially affected by Industrial Depression (1st Schedule, Part I and 1934 Act) in 1934 compared with 1911–1914 and 1924, and with England and Wales, etc.'; tabulations and correspondence with the Ministry of Health. 1934–1936.

[55] Bowley, *Housing and the State*; Yelling, *Slums and redevelopment*.

[56] Davidson, *Whitehall and the labour problem*, pp. 24–5.

(county borough, small town, rural, etc.) and, above all, region. The construction, and explanation, of the 'North/South divide' – why the northern areas of England and Wales consistently showed higher death rates than the registration counties south of the Wash – now became one of the central issues of mortality analysis.[57] By conceiving this in terms of 'The rule of Northern county borough maximum and Southern rural district minimum'[58], the GRO was taking Farr's 'Healthy District' model and stripping it of its local, political significance. Environment was now seen in terms of general causal factors operating across whole regions, rather than the local space over which political action needed to be mobilised to introduce sanitary reforms. In the hands of some inter-war medical statisticians this could lead to an emphasis on geography, climate and race as the prime causal factors in mortality differentials, thus undermining the possibility of effective local activity.[59]

The integration of the GRO into the Ministry of Health should not be seen, however, in an entirely negative light. As can be seen from Figure 7:1, which represents the pages of text in the *Statistical reviews* released prior to the Second World War, there was certainly no falling off in the overall statistical output of the Office. The only significant check to the expansion of the GRO's annual publication in the inter-war period was in the early 1930s, which coincided with an hiatus in the employment of a medical statistician in the GRO after Stevenson's death.

Similarly, the intervention of the Ministry of Health into the GRO's affairs brought a breakthrough in the organisation of the census. In January 1920 Vivian suggested in a memorandum to Morant that a compromise might be reached with the Treasury over the introduction of a permanent Census Act. A census could be taken every five years if authorised by an order in council, always providing that the Treasury was willing to pay.

---

[57] See, for example, GRO, *Supplement to the 75th annual report of the Registrar General: Part III Registration Summary Tables*, p. xxix; GRO, *Registrar General's statistical review for 1921*, pp. 12–13.

[58] GRO, *Registrar General's statistical review for 1922*, p. 14.

[59] Lewis-Faning, 'A study of the trend of mortality rates'. Lewis-Faning was a member of the MRC's Statistical Unit. For a contemporary critique of his arguments, see: Titmuss, *Birth, poverty and wealth*, pp. 52–3, 59–64, 76.

Figure 7.1 Pages of text in the Statistical Review

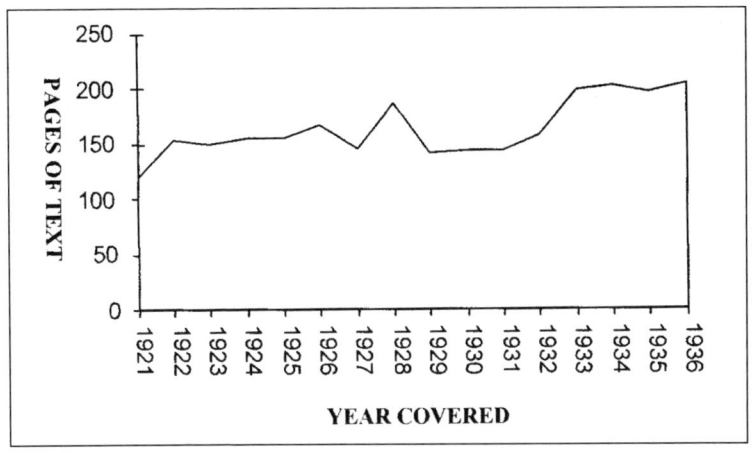

Source: Registrar General's Statistical Review for 1921 to 1936.

Morant sent it to Addison with the comment, 'Minister, ingenious and perhaps sufficient! Certainly difficult for the Chancellor of the Exchequer to rebut'.[60] Addison put up a paper to the Cabinet on the subject, whilst Vivian cleared his proposal with the government actuary. At the same time Winston Churchill wrote a Cabinet Memorandum from the War Office in which he maintained that a quinquennial census was 'essential to the proper solution of recruiting and manpower problems', and that if a new military service system was to be established, 'in a future emergency, it will be impossible to administer it until reasonably accurate vital statistics are available'.[61] The resulting 1920 Census Act provided for the taking of all future censuses, whether decennial or quinquennial. This facilitated the establishment of a permanent Census Branch within the GRO, although the subsequent re-assertion of Treasury control in the inter-war period prevented a true quinquennial census ever being taken.

---

[60] PRO: RG 19/49 Legislation: correspondence; memoranda; draft Census Bills, 1919–20, memo of Vivian to Morant 6 January 1920.
[61] Ibid., ff. 15, 20, 56.

'There are statisticians and statisticians' – medical statisticians in the inter-war period

Despite these developments, it could hardly be said that the GRO was now at the cutting edge of medical statistics. Its primary mission was to provide data rather than to initiate research, and certainly not to criticise other public authorities. In purely scientific terms the Office found itself at a disadvantage to other bodies, such as the Statistical Department of the MRC, because of its lack of advanced statistical skills and its narrow statutory remit.

In 1920 Greenwood noted in correspondence with Sir Walter Fletcher on developments in the GRO, there were 'statisticians and statisticians'.[62] By this he meant that the Registrar General's department had expertise in the collection and processing of data, as well as in the interpretation of their meaning, but that this did not require the specialist statistical skills at his disposal. The logic of Greenwood's chosen career path eventually led him to take over the MRC's Statistical Department on John Brownlee's death in 1927. At the same time he was appointed the first professor of epidemiology and vital statistics at the new London School of Hygiene and Tropical Medicine (LSHTM).[63] The academic research environment within which men such as Brownlee and Greenwood were located was conducive to a wider, experimental form of statistical work than could be pursued within a government department such as the Ministry of Health. According to an MRC memo of 1913:

> The object of the research is the extension of medical knowledge with the view of increasing our powers of preserving health and preventing and combating disease. But otherwise than that this is to be the guiding aim, the actual field of research is not limited and is to be wide enough to include, so far as may from time to time be found desirable, all researches bearing on health and disease, whether or not such researches have any direct or immediate bearing on any particular disease

[62] PRO: FD 5/91; letter from Greenwood to Fletcher, 17 June 1920.
[63] PRO: FD 2/13 Report of the Medical Research Council, 1926–1927, pp. 41–2.

or class of disease provided that they are judged to be useful in promoting the attainment of the above object.[64]

The desire to create a body that would allow its staff to approach research in such a flexible manner was the original reason for establishing the MRC outside the normal departmental structures of government.[65]

This greater flexibility can be seen in Brownlee's 1915 paper on the incidence of phthisis in the boot and shoe industry, which criticised the data provided by the GRO. Brownlee argued that the census classification 'shoemaker' was inconsistent, whilst the tendency for sick shoemakers to drop out of the trade, and for their deaths to be registered under other occupations, undermined the validity of the age specific mortality rates for the boot and shoe industry. He reworked the GRO's figures, and then used these in conjunction with information drawn from the National Union of Boot and Show Operatives and from the MOHs of Leicester and Northampton.[66] The deployment of non-registration data and advanced statistical methods was beyond the remit, and capability, of the GRO. Greenwood's posting to the LSHTM also allowed him more freedom to engage in new forms of scientific research. Whereas Farr and his successors had to depend on examining epidemics *post hoc* via the obscuring medium of the registration of causes of death, Greenwood could spend much of the 1930s at the LSHTM with his colleague W. W. C. Topley gaily infecting populations of mice with various diseases in order to reach a better understanding of epidemiology.[67]

In 1931, for example, the year of T. H. C. Stevenson's death, the staff of Greenwood's unit were completing a study of whooping-cough mortality; reporting on the Ministry of

[64] Quoted in Thomson, *Half a century of medical research. Volume 2*, pp. 3–4.
[65] Austoker, 'Walter Morley Fletcher and the origins of basic biomedical research', p. 24.
[66] PRO: FD 4/1 The first report of the special investigation committee upon the incidence of phthisis in relation to occupation. The boot and shoe industry.
[67] PRO: FD 2/14–24. For a summary of this work, see: PRO: FD 4/209 M. Greenwood, A. B. Hill, W. W. C. Topley and J. Wilson, 'Experimental epidemiology'.

Health's anthropological data on pre-school children; looking at the vitamin content of butter; studying birth rates in Wales and the south west division of England in the period 1860 to 1930; examining mortality from pulmonary tuberculosis in Wales; looking at the results of the insulin treatment of diabetes; analysing anthropological and nutritional data from Christ's Hospital, Horsham; considering the mathematical treatment of intelligence tests; looking at damage to the brain cortex and its effect on speech; studying the relationship between the different physical types of children and their liability to particular diseases, such as asthma and rheumatism; and completing work on minor epidemic illnesses in residential schools.[68] The staff of the Unit were encouraged to circulate their ideas on these subjects via the network of academic journals, rather than concentrating on the preparation of annual reports or reviews. In 1931, they had also given aid to the MRC's Committees on Industrial Pulmonary Disease and School Epidemics, and to the Industrial Health Research Board.[69] In the aftermath of the Second World War the Unit was to build up an international reputation by carrying out the first clinical trial undertaken on a rigorously statistical basis, that on the effects of streptomycin on tuberculosis, and providing the first statistical proof of the link between smoking and lung cancer.[70] The GRO could not possibly hope to match the range and innovative nature of this work but it could draw on it in a derivative fashion by quoting some of the resulting articles in its own publications.[71]

The intellectual subordination of the GRO was given official form by a 'concordat' signed between the MRC and the Ministry of Health in January 1924. This arose out of Fletcher's

---

[68] PRO: FD 2/18 Report of the Medical Research Council, 1931–1932, pp. 113–20.

[69] PRO: FD 2/18, p. 113.

[70] Matthews, *Quantification and the quest for medical certainty*, pp. 115–40; Thomson, *Half a century of medical research,. Volume 2*, pp. 238–9; Doll and Hill, 'Smoking and carcinoma of the lung'; Doll and Hill, 'A study of the aetiology of carcinoma of the lung'; Doll and Hill, 'Lung cancer and other causes of death in relation to smoking'. The Unit also produced Austin Bradford Hill's classic *Principles of medical statistics*. See also, Higgs, 'Medical statistics, patronage and the state'.

[71] For example: GRO, *Registrar General's Statistical Review for 1923*, p. 34; GRO, *Registrar General's Statistical Review for 1925*, p. 46.

belief at the MRC that all medical research in the UK should be brought under the control of his Council. This empire building led to 'turf wars' with the Ministry, which was pursuing its own research agenda. At the suggestion of the Treasury, Fletcher and Newman at Health agreed to a general cessation of hostilities and a carving up of the field of medicine. The Ministry was in future to concentrate on 'applied research' relating to clinical problems, whilst the MRC was given the task of initiating and organising all new research.[72] In the field of statistics the Ministry (which expressly included the GRO) was, 'To survey by statistical or other means existing states of national (and international) health and environment, both absolutely and in relation to past history.' The MRC's sphere was to be, 'Medical research by statistical and other methods (primarily for the development of new methods of statistical enquiry)'.[73] A separate undated memo, 'Co-ordination of functions of the Ministry of Health and of the Medical Research Council', also envisaged that the central functions of the two departments would lie in different fields. The job of the Ministry was administration, to which 'research' in its various kinds would be ancillary. The function of the Council was research work as such, not combined with any administrative responsibilities. Under the heading of statistics it was re-affirmed that:

> The Ministry must chiefly use established methods for statistical inquiry. The MRC aims at improving methods of inquiry or finding new mathematical weapons. The MRC has also to bring statistics in new ways to assist laboratory and clinical research.[74]

The GRO was thus debarred from the development of the new advanced statistical techniques, and increasingly forced back into the servicing of policy requirements.

---

[72] Thomson, *Half a century of medical research. Volume 1,* p. 72; Austoker, 'Walter Morley Fletcher and the origins of basic biomedical research policy', pp. 25–6.
[73] PRO: FD 1/1374 Ministry of Health and MRC – policy. Pt 1; concordat of 22 January 1924.
[74] Ibid., memorandum entitled 'Co-ordination of functions of the MH and of the Medical Research Council'.

The centre for statistical guidance in medical matters in government in the inter-war period was not, therefore, the GRO but the Statistical Committee of the MRC, of which Major Greenwood was chairman from its inception in 1921 till 1948. The creation of this body was part of Walter Fletcher's strategy of fostering small select committees, the members of which would form a scientific elite exerting control over a field of research through the selective support of certain individuals and projects.[75] Fletcher used the Statistical Committee to co-ordinate statistical work within the MRC, and thus to subordinate the ineffectual John Brownlee to Greenwood.[76] The Committee offered help to other sections of the MRC and government departments, and was responsible for editing and approving statistical texts for publication in the MRC's reports series. It originally had its own statistical staff but these were merged with those of the Statistical Department when Greenwood moved to the LSHTM in 1927.[77] In time the Statistical Committee grew into an inter-departmental body with representatives from the MRC, Ministry of Health, Government Actuaries Department, and, from 1927, Stevenson from the GRO.[78] In practice, however, the main work of the Committee devolved on a small group of Greenwood's friends and protégés.[79] Nevertheless, Fletcher saw the body as a national centre for all statistical work in medicine.[80] Greenwood agreed with Newman at the end of 1926 that there should be complete interchangeability between the work done for his committee and that for the Ministry of Health.[81] This

---

[75] Austoker, 'Walter Morley Fletcher and the origins of basic biomedical research policy', p. 27.
[76] Thomson, , *Half a century of medical research. Volume 1*, pp. 114–5; Austoker and Bryder, 'The National institute for Medical Research and related activities of the MRC', p. 51.
[77] Ibid., p. 52.
[78] PRO: FD 2/13 Report of the Medical Research Council, 1926–1927, p. 41.
[79] PRO: FD 1/7114 Committee on Industrial Health Statistics, Industrial Fatigue Research Board; memo from Fletcher to MRC Council, 20 February 1925; FD 1/7108 Statistical Committee: Vol. II, 1927–9; 'Memorandum on organisation'; letter from Greenwood to Landsborough Thomson, 21 December 1927.
[80] PRO: FD 1/7107 Statistical Committee: establishment and correspondence; letter from Fletcher to Newman, 8 June 1925.
[81] Ibid.; extract from Committee minutes of 21 December 1926.

body fulfilled, therefore, some of the function of the general statistical council that had been envisaged at the establishment of the Ministry of Health.

The intellectual eclipse of the GRO's medical officers by the MRC Statistical Committee became a formal subordination on the death of Stevenson in September 1931. Rather than replacing him immediately, it was agreed between the Ministry of Health, the MRC and the LSHTM, that Greenwood should act as honorary medical advisor to the Registrar General in his role as chairman of the MRC Committee. A small sub-committee of the latter, made up of Greenwood, the ubiquitous Yule, and Greenwood's protégé Leon Isserlis, was established, 'to consider any matters of statistical principle upon which the Registrar General may seek advice'.[82] Greenwood's apparent self-sacrifice was explained, in part, by his fear that Newman at Health was about to cut off the Ministry's grant for the Statistical Unit at the LSHTM. However, as he explained to Fletcher, he also saw the possibility of bringing, 'the whole scientific control of the public vital statistics under the MRC ...'.[83]

This situation continued until late in 1933 when Dr Percy Stocks was appointed as 'medical statistical officer' at the GRO under Derrick, the latter having been promoted to assistant Registrar General.[84] Stocks had been educated at King's College, Cambridge, where he took the Natural Science Tripos, Victoria University, Manchester, and the School of Tropical Medicine, Liverpool. He started out his career as a house physician at Manchester Royal Infirmary, and became a temporary lieutenant in the Royal Army Medical Corp during the First World War, where he worked in general practice and bacteriology. He was assistant school medical officer in Bristol in the period 1918 to 1921, before going on to become reader in medical statistics at the Galton Eugenics Laboratory under Pearson, from 1921 to 1933. How far Stocks was a eugenicist is a moot point. In the years immediately preceding his appointment to the GRO he

---

[82] PRO: FD 2/18 Report of the Medical Research Council, 1931–32, p. 112.
[83] PRO: FD 1/7109 Statistical Committee: Vol. III, 1929–1931; letter from Greenwood to Fletcher, 25 September 1931.
[84] HMSO, *Imperial Calendar 1934* (London, 1934), p. 190.

was a regular contributor to Pearson's *Annals of Eugenics* but his work was mainly epidemiological rather than overtly eugenic.[85] Stocks joined the MRC Statistical Committee late in 1934, and continued working at the GRO until 1950, although in the 1940s he was increasingly drawn into the work of the World Health Organisation.[86]

The advent of Stocks was associated not only with the rekindling of interest in social class as an explanatory variable within the GRO's publications but also with an increasing statistical sophistication. In the *Registrar General's statistical review for 1934*, published in 1936, Stocks introduced the use of correlation coefficients and regression analysis.[87] The same *Review* also noted the introduction of local areal comparability factors (AFCs) to allow standardisation between localities.[88] The *Review* published in 1938 contained a discussion of the need to introduce measures of statistical significance in those cases where the number of cases of a particular cause of death was under 20. Stocks showed that the standard error decreased as a percentage of the number of deaths by cause as the number of cases increased – plus or minus 32 per cent with 10 cases, and plus or minus 3 per cent at 1,000.[89] The probability statistics of Galton, Pearson and their successors, had finally arrived in the GRO, infusing the older statistical tradition of Farr with a new rigour and analytical power. The link between the GRO's reviving interest in social class, and Stock's work within a eugenic framework, also cannot be discounted.

---

[85] Stocks, 'Fresh evidence on the inheritance factor in tuberculosis'; Stocks, 'Infant mortality in the metropolitan boroughs in relation to occupation'; Stocks, 'A study of the epidemiology of measles'; Stocks, 'A biometric investigation of twins and their brothers and sisters'; Stocks, 'The distribution of cancer and tuberculosis mortality in England and Wales'; Stocks, 'On the spread of small-pox in partially vaccinated communities'.

[86] PRO: FD 1/7110 MRC Statistics Committee: Report, including anaemia and measles, 1933–1936, extract of the Committee meeting of 26 October 1934; *Who was who, 1971–1980*, pp. 760–1; GRO, *Registrar General's statistical review for 1951: Text*, p. 275.

[87] GRO, *Registrar General's statistical review for 1934*, pp. 150–5.

[88] Ibid., pp. 4–8.

[89] GRO, *Registrar General's statistical review for 1936*, pp. 9–15.

Statistics and national planning in the Welfare State

It was, however, the Second World War, and the events leading up to it, that really transformed the role of the GRO. The threat of a European conflict reactivated interest in both national mobilisation and population trends. Also, health care became a right of national citizenship which could be claimed anywhere within the nation, rather than linked to local charitable provision, or payment into a specific scheme attached to a designated hospital, or medical practice, as with the national insurance system. New citizens' rights meant that new fields of data gathering opened up for the purpose of informing central administrative policy. This led to a shift in the GRO's orientation from a provider solely of mortality statistics to the generation of data on national morbidity rates.

The preparations for a new national registration system began at least as early as 1935 because, in Sylvanus Vivian's words, of 'the recent inter-national disturbances'.[90] In that year the Registrar General was chairing a sub-committee of the Committee on Imperial Defence on the subject, and a draft National Service Bill had been drawn up.[91] With the outbreak of war in 1939 the enumeration machinery was activated to compile the national register. The population figures produced were seen by the GRO as the nearest thing to a census likely to be taken in war conditions, and as such they were first circulated for official use and then published in 1944.[92] As in the First World War, the register was used for rationing purposes, and for the deployment of labour in the military and other essential services, and also for the issue of identity cards. The onset of the

[90] PRO: RG 28/31 Committee of Imperial Defence: Registrar General's sub-committee on National Registration: correspondence and papers 1935–39; letter of Vivian to E A Armstrong, 17 December 1935.
[91] Ibid.; memorandum on 'Draft Bills in Appendix C and Appendix E to document 350–B'. Both the official *Guide to census reports,* and Nissel mistakenly give the impression that the preparations for the national registration system only began in 1938: Office of Population Censuses and Surveys and the General Register Office, Edinburgh, *Guide to census reports,* p. 26; Nissel, *People count,* p. 75.
[92] RG 26/6 National Register 1939: preparation in the event of war ... 1938–1941; draft circular to clerks of local authorities, 1940; *National registration. Statistics of population on 29 September 1939 by age, sex and marital condition: report and tables* (London, 1944).

Cold War meant that national registration, and national service which it underpinned, was not abolished after the defeat of Germany. However, despite its use for the purposes of organising national service and consultation by the security services and police, the national registration system and the issuing of identity cards, was wound up in 1952.[93] This was partly on the grounds of expense (the system cost £500,000 per annum) but also because the carrying of identity cards, and the need to register changes of address, was widely regarded as 'unnecessary and oppressive'.[94] On the abolition of national registration, the register was then used to prepare the NHS Central Register (NHSCR), which was subsequently kept up to date by local registrars of births and deaths, and by family practitioner committees. This was used to monitor the transfer of patients between general practitioners; for the issue of NHS numbers; and to notify family practitioner committees when people no longer needed health care because they had died, or moved abroad.[95] By 1951 the GRO was employing 870 staff, out of a total of 1,486, in departments dealing with various aspects of registration and the provision of direct services to the public.[96]

The belief that the British population was in decline in a period of renewed international tension also led to the passing of the 1938 Population (Statistics) Act.[97] This led to the introduction on 1 July 1938 of additional questions to be asked at the time of birth registration, including: the age of the mother; and, for legitimate births, the parents' date of marriage, and the number of live and still births to the mother within marriage. The GRO subsequently used this extra data in its published discussions of marital fertility. The civil tables in the Office's *Statistical review for 1938,* for example, covered numerous facets of the subject, including: the ages of mothers; legitimate and

---

[93] PRO: CAB 134/907 Home Affairs Committee Memorandum H.A.(51)23.
[94] PRO: RG 28/201 Discontinuance of the National Registration Act, 1939: proceedings in Parliament.
[95] Nissel, *People count,* pp. 75–77.
[96] PRO: RG 20/50 General Register Office and Office of Population Censuses and Surveys, Establishment and Accounts Division: Correspondence and Papers.
[97] For the events surrounding the passage of the Act, see: Soloway, *Demography and degeneration,* pp. 226–58.

illegitimate fertility; sex ratios and regional distributions of births; multiple births; an analysis of previous children of all marriages; the duration of marriages; first maternities; family size according to marriage duration; infertility; and so on.[98] Public concern over the subject of population decline led to the appointment of a Royal Commission on Population in 1943 with the job of examining population trends in Great Britain, investigating the causes of those trends, and considering their probable consequences. It was also tasked with making suggestion as to what could be done about the problems that low population growth would throw up.[99] The Commission, which reported in 1949, drew upon the assistance of the GRO for a number of special enquiries.[100] Anxiety over national population trends in an age of international conflict thus maintained the relevance of the GRO's statistical functions.

It was above all, however, in the field of health care, both during and after the War, that the GRO found a new outlet for its activities and a new role in central planning. Fear of massive civilian casualties on the outbreak of war, and the need to relocate large numbers of evacuees, led to the establishment of a national Emergency Hospital Service (EHS).[101] Percy Stocks soon realised that such centrally planned arrangements required data on morbidity to function efficiently. In response, he first organised a one-in-five survey of patients being admitted to the EHS to discover their illnesses and patient histories.[102] In 1943 he negotiated with the newly established Wartime Social Survey to undertake a 'Sample Survey of Sickness' to discover the nature and severity of morbidity in the population.[103] Significantly, the

[98] GRO, *Registrar General's statistical review for 1938* : Civil tables, pp. 108–207.
[99] GRO, *Registrar General's statistical review for 1938 and 1939*, p. 179. For the events leading up to the appointment of the Commission, see: Soloway, *Demography and degeneration*, pp. 312–43.
[100] The GRO's own records on the Commission comprise a whole class of its records at the PRO: RG 24.
[101] Dunn (ed.), *The Emergency Medical Services*.
[102] PRO: RG 26/83 Hospital in-patients summary: policy 1947–1951; 'Explanatory memorandum on the use of hospital in-patient summary'.
[103] PRO: RG 26/24 Sample Survey of Sickness 1943–45 conducted by the Wartime Social Survey: initial planning and development; letter from Taylor to Stocks of 24 July 1943. For the history of the Government Social Survey, see: Whitehead, 'The Government Social Survey'; Moss, *The Government Social Survey*.

latter was classified according to the amount of time someone was incapacitated for work.[104] By July 1944, Stocks was advising the Ministry's chief medical officer that the survey would enable it to measure the 'total loss of working time and total service by doctors ...'. Similarly, the survey would help in 'guiding policy in the building up of health services ...'.[105] This new form of data collection proved to be better than the existing system for notifying the Ministry of Health of notifiable diseases, which it subsequently replaced. The Survey revealed that the only notifiable diseases for which notification could be regarded as fairly complete were acute poliomyelitis, cerebro-spinal fever, diptheria and scarlet fever. Respiratory tuberculosis came next with probably nine tenths of active cases notified.[106] The survey continued to be taken until 1952, and its findings were published regularly in the GRO's *Quarterly returns.*[107] The Survey of Sickness also provided the Ministry of National Insurance with a means of tracking the likely fluctuations in the take-up of sickness benefits, as in the case of the flu epidemic of 1951.[108]

As the war drew to a close, the GRO consolidated its new field of responsibility. In May 1945 a meeting was held at the Ministry of Health to discuss the subject of national morbidity statistics with participants from Health, the Industrial Health Research Board, the Ministry of Education, the Ministry of National Insurance, the Wartime Social Survey, the MRC Statistical Unit, and the GRO. Stocks circulated participants with a memorandum on the subject, which assumed that the future collection of such data on a national basis would be carried out centrally by the GRO.[109] In December 1945 an Anglo-US-Canadian Working Party, including Stocks, began the task of drafting a single international classification which

---

[104] PRO: RG 26/24 Sample Survey of Sickness 1943–45 conducted by the Wartime Social Survey: memorandum on 'MRC classification of diseases and injuries'.
[105] Ibid.; memorandum from Stocks to the chief medical officer of 3 July 1944.
[106] GRO, *Registrar General's statistical review for 1946–1947: Medical*, p. 3.
[107] Nissel, *People count*, pp. 87–9, 117; Logan and Brooke, *The survey of sickness.*
[108] GRO, *Registrar General's statistical review for 1951: Text* , p. 4.
[109] PRO: RG 26/398 National morbidity statistics: general policy.

could be used for both morbidity and mortality statistics. After amendment, this was accepted in 1948 as the new International List of the Causes of Death by the World Health Organisation. With the prospect that a satisfactory classification for morbidity statistics would soon be available, a Working Party on Hospital Records was set up in 1945 to consider whether collecting statistics by sampling the records of patients admitted to beds in the EMS hospitals could be extended.[110] The ubiquitous Stocks drafted a summary form for use in collecting such statistics of hospital in-patients, seeing such information as 'essential for the purposes of efficient direction of a national hospital scheme, answering parliamentary questions and guidance in the matter of reducing waiting lists and provision of new hospitals.'[111] At the end of 1947, the Working Party on Hospital Records recommended to the Ministry of Health that a pilot survey should be started in selected hospitals.[112] This proved a success and since 1949 the GRO, and its successors, have run a Hospital In-Patient Enquiry (HIPE) along the lines originally suggested by Stocks for the purpose of allocating resources within the NHS.[113] It should be noted, however, that there was nothing inevitable about this development since bodies such as the Advisory Committee on Scientific Policy were considering the establishment of a more general Medical Statistics Service at this date.[114]

By the early 1950s the GRO was undertaking a number of other surveys for the Ministry of Health that had a direct bearing on the provision of services in the NHS and the health of citizens. At the beginning of 1947 the Office took over responsibility from the Radium Commission for the central collection and analysis of records relating to registered cases of cancer in England and Wales.[115] In 1950 the GRO superseded the

---

[110] GRO, *Registrar General's statistical review for 1946–1947: Medical*, pp. 1–2.
[111] PRO: RG 26/83; memorandum of 14 April 1947 entitled 'Information to be obtained from National Statistics of Hospital Inpatients'.
[112] GRO, *Registrar General's statistical review for 1946–1947: Medical*, p. 2.
[113] Nissel, *People count*, p. 116.
[114] PRO: RG 26/398; Advisory Council on Scientific Policy; Medical Statistical Service, note by the chairman, item 21E.
[115] GRO, *Registrar General's statistical review for 1946–1947: Medical*, p. 3.

Ministry as the body tasked with the analysis of data generated by the mass radiography programme, which had been established during the Second World War as a means of combating tuberculosis.[116] The following year a pilot scheme was initiated in which ten general practitioners started recording details of the ailments of their patients for subsequent statistical analysis by the GRO. This was to grow into a full-blown survey of the services provided by GPs in the NHS.[117] At the same time the Office was undertaking a investigation into rubella, and mounting a mental hospital enquiry for the purpose of meeting 'the special legal and administrative problems raised by mental cases'.[118] The GRO's new role as a generator of data on morbidity also fuelled new publishing ventures. From 1947 the Office began publishing a free-standing series of 'Studies on medical and population subjects'. The second of these was a volume by Stocks on *Sickness in the population of England and Wales in 1944–1947,* followed by a third on cancer registration.[119] From the early 1950s the GRO was regularly publishing supplements to the *Statistical review* on general morbidity and hospital in-patient statistics.[120] At this date the Office was employing 263 staff in its various professional statistical branches.[121]

In a sense the GRO had come full circle – just as the generation of data on illness had, in part, underlain the creation of the GRO's Statistical Department in the 1830s, so now its work on the morbidity of the population underpinned the emerging NHS. But the citizen of the early Victorian period was being offered information to make his own decisions with regard to insuring himself and his family privately against the effects of ill-health and death, or in order to help him force local sanitary authorities to undertake their obligations. The role of the GRO in the twentieth century, however, was to provide data for the

---

[116] PRO: RG 26/170 Mass Radiography 1945 to 1970: Policy: transfer of statistical processing from Ministry of Health, analysis of data 1945; 1949–65.
[117] GRO, *Registrar General's statistical review for 1951: Text,* p. 2.
[118] PRO: RG 26/398; Ministry of Health memorandum 56B.
[119] GRO, *Registrar General's statistical review for 1946–1947: Medical,* p. 3.
[120] GRO, *Registrar General's statistical review for 1950: Medical,* p. 1.
[121] PRO: RG 20/50.

central planning of health care by the nation state. As the *Statistical review* published in 1953 put it:

> The increasing responsibility of the central government for provision in the field of medical treatment as well as prevention demands an increasing knowledge by the government departments concerned of the state of health or sickness of the people and the changing demands likely to be imposed on available resources for maintaining and restoring health.[122]

The reorientation of the GRO from Victorian liberalism to twentieth-century christian/social democracy had been completed.

---

[122] GRO, *Registrar General's statistical review for 1948–49: Medical*, p. 1.

8

Conclusions

Certain general conclusions can be drawn from this organisational history of the GRO. The first relates to the role of contingency in history. Although the early nineteenth century saw the general development of state statistical bureaux across Europe, a reflection of the creation of the modern nation-state[1], it was not inevitable that the GRO would fulfil this function in England. At its inception the GRO was not a statistical bureau but the centre of a system for the recording of property rights. In truth the latter always dominated the institution's staffing and operations throughout the period under consideration here. If the Office established and expanded its Statistical Department over time, this was because of the opportunism of men such as Thomas Lister, George Graham, Bernard Mallet, William Farr, T. H. C. Stevenson, and Percy Stocks. Although they worked within the context of increasing state demands for medical, social and actuarial statistics, it was their initiative and talent that led to this demand being met by the GRO. The fate of the Office under a weak man such as Brydges Henniker, or a Registrar General with limited aims, such as Sylvanus Vivian, reveals the importance of such factors.

What was crucial here, however, was the initiative of the Registrar General, rather than that of his statistical subordinate, the Superintendent of Statistics. Without the active support of a strong head of the Office, men of great talent, such as William Ogle and John Tatham towards the end of the nineteenth century and Stevenson in the 1920s, might find their opportunities for statistical expansion limited. The key to statistical innovation was, therefore, fruitful co-operation between

---

[1] Higgs, *The Information State in England,* pp. 19–21.

administrator and statistician, as in the case of Graham and Farr, and Mallet and Stevenson. A possible exception to this pattern was the ability of Stocks to carve out a new statistical territory for the GRO during and after the Second World War. But this probably reflected the wholly novel conditions of 'War Socialism'. This realisation helps to rescue the reputations of some worthy administrative civil servants from the condescension of academic posterity.

A consideration of the importance of administrative factors in the history of the GRO's development also raises the question of the centrality of intellectual arguments to this story. Nothing has been said in the present work that should be taken to imply that ideas and scientific debates were not important to an understanding of the Office's statistical activities. However, the actual impact of currents of thought on the GRO's published output is not always easy to determine, and that of some intellectual movements, such as eugenics, may not have been quite what some historians have believed. Certainly factors such as the competence of the Registrar General, the structure of the Civil Service, and the relationships between Whitehall departments, have to be given equal weight. Similarly, the role of eugenics in the development of the GRO might have as much to do with changes in computational and data management technologies, as in the provision of an intellectual stimulus.

Life, death and statistics has attempted to place the history of the GRO within the context of two interlinked processes within modern state formation in England. The first is the changing nature of citizenship as the English polity changed from one based on classical liberal principles in the nineteenth century to one based on social democracy in the twentieth. Statistical production in the GRO in the High Victorian period was about providing information for individuals to make decisions about their lives in a market economy, to develop medical science, or to facilitate local democracy. By the early 1950s statistical production was increasingly about providing intelligence to underpin the provision of good and services to citizens by the central state. This process led in the twentieth century to a restriction of the GRO's freedom of action, via its increasing subordination to government departments such as the Ministry of Health and the Treasury.[2]

The organisational history of the GRO helps to explain the form and scope of the institution's published output. The GRO was originally a place for the recording of births, marriages and deaths for the purpose of determining the lines of descent that affected property rights. This explains why it did not record morbidity or stillbirths, and why registration was not in effect compulsory until the early 1870s. The origins of the Office's statistical function in actuarial work, and its precarious beginnings, explains many of the idiosyncratic features of its early, published output. On the other hand, unlike many state statistical bureaux on the Continent, the statistical work of the GRO was based on an impressive database of vital records. Such issues have important implications for the historian's understanding of the form and reliability of the GRO's tables, and thus of the historiography based on that material.

The expansion and contraction of the GRO's published output, and its changing orientation, is also explicable in organisational, and well as intellectual, terms. The origins of the Office's statistical work lay in the initiative of Lister and Chadwick, and in the unusually independent position of the GRO under the 1836 Registration Act. The period of innovation and growth up to 1880 reflected the fruitful collaboration between Graham and Farr, the latter providing intellectual input and the former administrative flair. This enabled the GRO to take advantage of the increasing demand by government for public health data. The relative stagnation of the last twenty years of the nineteenth century can be exaggerated but seem to have reflected, in part, the Office's staffing difficulties and the weakness of Brydges Henniker. The recovery of the institution's statistical output in the early twentieth century may have been partly due to the controversies over eugenics, although the GRO was not wholly in the opposition camp. However, the general expansiveness of the Civil Service under 'New

---

[2] This is part of the general argument of Higgs, *The Information State in England.*

Liberalism', better senior management, and new forms of data processing, all played their part. Lastly, the GRO's organisational subordination to the Ministry of Health after the First World War helps to explain the increasing anonymity and limitations of its statistical output in the interwar period. But the expansion of the central state's health care apparatus during and after the Second World War also led to new departures in the Office's statistical work.

As all serious archival researchers know, one cannot say exhaustively why historical artefacts appear in a certain form and at a certain date. However, one does not have to fall back on vague postmodernist scepticism about the slipperiness of meaning, and the gap between the present and the past. The way in which bureaucratic and organisational forces help to sculpt the form and scope of the records on which historians base there work is explicable to us because we are, thankfully only in part, organisational creatures ourselves. Anyone struggling to achieve something within an institution will recognise the importance of effective and supportive management, the scope to be creative and take initiative, and of organisational culture. This should help us to empathise with the good fortune of a William Farr, and with the sorry fate of his successor, T. H. C. Stevenson.

Appendix 1

Registrar general's annual reports, Statistical reviews, and Decennial supplements, 1838–1951

The following listing is not an attempt to produce an exhaustive catalogue of the published output of the GRO over the period 1838 to 1951. From its inception the Office always produced material other than its main run of *Annual reports, Decennial supplements*, and *Statistical reviews*, such as its quarterly and weekly reports. The GRO also introduced a number of new series after the Second World War. The listing does, however, give a generally complete summary of the longest series of medical reports released by the GRO in these years.

Until 1920 the *Annual report of the Registrar General of births, marriages and deaths for England and Wales* (hereafter ARRG) was published as a Parliamentary Paper but separate *Reports* were also produced and circulated to the copyright libraries and elsewhere. In the 1840s and early 1850s these two series contained differing material, although after that for 1854 they were always identical. Indeed, many of the later volumes held separately by the copyright libraries are merely the Parliamentary Paper versions. In the following listing, details of the two series are given separately until the report for 1854, and variations between the two are briefly noted. For the *Annual reports* for 1855 onwards, only details of the Parliamentary Paper series are given.

After 1920 the *Annual report* was replaced by the *Statistical review*, which was never issued as a Parliamentary Paper. Prior to 1920, the *Decennial supplements to the annual reports* were also produced in two series but were always identical. After 1920 they ceased to be produced as Parliamentary Papers. The production and numbering of the *Decennial supplements* became very irregular in the twentieth century under the double impact war and financial crises.

Annual reports of the Registrar General

Year ending 30 June 1838
1st ARRG for year ending 30 June 1838 (London,1839 – dated May 18, 1839)
1st ARRG for year ending 30 June 1838 (dated May 18, 1839), PP 1839, XVI [187.]

Year ending 30 June 1839
2nd ARRG for year ending 30 June 1839 (London, 1840 – dated June 1840)
2nd ARRG for year ending 30 June 1839 (dated June 1840), PP 1840, XVII [263.] [276.]

Year ending 30 June 1840
3rd ARRG for year ending 30 June 1840 (London, 1841 – dated June 1841)
3rd ARRG for year ending 30 June 1840 (dated June 1841), PP 1841, 2nd session, VI [345.] [348.]

Year ending 30 June 1841
4th ARRG for year ending 30 June 1841 (London, 1842 – dated 8 August 1842)
4th ARRG for year ending 30 June 1841 (dated 8 August 1842), PP 1842, XIX [423.]

1841
5th ARRG for 1841 (2nd ed. 'revised and corrected', London, 1843– dated 14 Aug 1843)
5th ARRG for 1841 (dated 14 Aug 1843), PP 1843, XXI [515.]
The non-Parliamentary Paper has additional tables in both Graham's report and Farr's *Letter*.

1842
6th ARRG for 1842 (London, 1845 – dated 10 Aug 1844)
6th ARRG for 1842 (dated 10 Aug 1844), PP 1844, XIX [540.]
The non-Parliamentary Paper has additional tables in both Graham's report and Farr's *Letter*

1843 and 1844
7th ARRG for 1843 and 1844 (London, 1846 – dated Dec 29 1845)
7th ARRG for 1843 and 1844 (dated Dec 29 1845), PP 1846, XIX [727.]
The Parliamentary Paper has a circular to doctors, and extra tables on the population of Kent. The non-Parliamentary Paper has extra notes to some extra tables,

Appendix 1

copies of circulars to doctors and coroners on classification of causes of death, and life tables.

1845

8th ARRG for 1845 (London, 1849 – dated 25 March 1847)
8th ARRG for 1845 (dated 25 March 1847), PP 1847–8, XXV [967.]
Both versions have tables not in the other.

1846

9th ARRG for 1846 (London, 1849 – dated 1 Aug 1848)
9th ARRG for 1846 (dated 1 Aug 1848), PP 1847–8, XXV [996.]
Appendix to the 9th ARRG for 1846 (no date), PP 1849, XXI [1087.]
The non-Parliamentary Paper contains extra meteorlogical data and tables, and an index to the report.

1849

10th ARRG for 1847 (London, 1852 – dated 1 Dec 1850)
10th ARRG for 1847 (dated 24 July 1849), PP 1849, XXI [1113.]
The Parliamentary Paper merely gives some abstracts of births, marriages and deaths. The full report and additional abstracts are in the non-Parliamentary Paper.

1848

11th ARRG for 1848 (London, 1852 – dated 1 Dec 1851)
11th ARRG for 1848 (dated 24 July 1850), PP 1850, XX [1255.]
The Parliamentary Paper merely gives some abstracts of births, marriages and deaths. The full report and additional abstracts are in the non-Parliamentary Paper.

1849

12th ARRG for 1849 (London, 1853 – dated 10 Jan 1853)
12th ARRG for 1849 (dated 24 July 1851), PP 1851, XXII [1416.]
The Parliamentary Paper merely gives some abstracts of births, marriages and deaths. The full report and additional abstracts are in the non-Parliamentary Paper.

1850

13th ARRG for 1850 (London, 1854 – dated Aug 21, 1854)
13th ARRG for 1850 (dated 24 June 1852), PP 1852, XVIII [1520]
The Parliamentary Paper merely gives some abstracts of births, marriages and deaths. The full report and additional abstracts are in the non-Parliamentary Paper.

1851

14th ARRG for 1851 (London, 1855 – dated 23 May, 1855)

14th ARRG for 1851 (dated 24 June 1853), PP 1852–53 XL [1665.]
The Parliamentary Paper merely gives some abstracts of births, marriages and deaths. The full report and additional abstracts are in the non-Parliamentary Paper.

1852
15th ARRG for 1852 (London, 1855 dated 30 August 1855)
15th ARRG for 1852 (dated 24 June 1854), PP 1854 XIX [1823.]
The Parliamentary Paper merely gives some abstracts of births, marriages and deaths. The full report and additional abstracts are in the non-Parliamentary Paper.

1853
16th ARRG for 1853 (London, 1856 – dated 10 Dec, 1855).
16th ARRG for 1853 (dated 24 June 1855), PP 1854–5 XV [1970.]
The Parliamentary Paper merely gives some abstracts of births, marriages and deaths. The full report and additional abstracts are in the non-Parliamentary Paper.

1854
17th ARRG for 1854 (London, 1856 – dated 30 Aug, 1856).
17th ARRG for 1854 (dated 24 May 1856), PP 1856 XVIII [2092.]
The Parliamentary Paper merely gives some abstracts of births, marriages and deaths. The full report and additional abstracts are in the non-Parliamentary Paper.

1855 *18th ARRG for 1855* (dated 10 June, 1857), PP 1857 sess. 2, XXII [2260.]
1856 *19th ARRG for 1856* (dated 10 June, 1858), PP 1857–58 XXIII [2431.]
1857 *20th ARRG for 1857* (dated 3 May, 1859), PP 1859 sess. 2 XII [2559.]
1858 *21st ARRG for 1858* (dated 30 June, 1860), PP 1860 XXIX [2712.]
1859 *22nd ARRG for 1859* (dated 30 May, 1861), PP 1861 XVIII [2897.]
1860 *23rd ARRG for 1860* (dated 15 April, 1862), PP 1862 XVII [2977.]
1861 *24th ARRG for 1861* (dated 28 March, 1863), PP 1863 XIV [3124.]
1862 *25th ARRG for 1862* (dated 1 June, 1864), PP 1864 XVII [3415.]
1863 *26th ARRG for 1863* (dated 31 May 1865), PP 1865 XIV [3562.]
1864 *27th ARRG for 1864* (dated 30 April 1866), PP 1866 XIX [3712.]
1865 *28th ARRG for 1865* (dated 25 June 1867), PP 1867 XVII [3913.]
1866 *29th ARRG for 1866* (dated 31 March 1868), PP 1867–68 XIX [4006.]
1867 *30th ARRG for 1867* (dated 31 March, 1869), PP 1868–69 XVI [4146.]
1868 *31st ARRG for 1868* (dated 31 March, 1870), PP 1870 XVI [c. 97]
1869 *32nd ARRG for 1869* (dated 1 July, 1871), PP 1871 XV [c.453]
1870 *33rd ARRG for 1870* (dated 26 June, 1872), PP 1872 XVII [c.667]
1871 *34th ARRG for 1871* (dated 26 May, 1873), PP 1873 XX [c.806]
1872 *35th ARRG for 1872* (dated 7 July, 1874), PP 1875 XVIII Pt 1 [c. 1155]

Appendix 1   225

1873 *36th ARRG for 1873* (dated 11 August, 1875), PP 1875 XVIII Pt 1 [c. 1155]
1874 *37th ARRG for 1874* (dated 31 July, 1876), PP 1876 XVIII [c.1581]
1875 *38th ARRG for 1875* (dated 18 June, 1877), PP 1877 XXV [c.1786]
1876 *39th ARRG for 1876* (dated 30 March, 1878), PP1878 XXII [c. 2075]
1877 *40th ARRG for 1877* (dated 31 March, 1879), PP 1878–79 XIX [c.2276]
1878 *41st ARRG for 1878* (dated 31 March, 1880), PP 1880 XVI [c.2568]
1879 *42nd ARRG for 1879* (dated 31 March, 1881), PP 1881 XXVII [c.2907]
1880 *43rd ARRG for 1880* (dated 31 March, 1882), PP 1882 XIX [c.3208]
1881 *44th ARRG for 1881* (dated 31 March, 1883), PP 1883 XX [c.3620]
1882 *45th ARRG for 1882* (dated 31 March, 1884), PP 1884 XX [c.4009]
1883 *46th ARRG for 1883* (dated 31 March, 1885), PP 1884–85 XVII [c.4424]
1884 *47th ARRG for 1884* (dated 31 March, 1886), PP 1886 XVII [c.4722]
1885 *48th ARRG for 1885* (dated 21 December, 1886), PP 1886 XVII [c.4896]
1886 *49th ARRG for 1886* (dated 30 November, 1887), PP 1887 XXIII [c.5138]
1887 *50th ARRG for 1887* (dated 31 October, 1888), PP 1888 XXX [c.5590]
1888 *51st ARRG for 1888* (dated 30 November, 1889), PP 1889 XXV [c.5846]
1889 *52nd ARRG for 1889* (dated 29 November, 1890), PP 1890 XXIV [c.6170]
1890 *53rd ARRG for 1890* (dated 26 November, 1891), PP 1890–91 XXIII [c.6478]
1891 *54th ARRG for 1891* (dated 30 November, 1892), PP 1892 XXIV [c.6841]
1892 *55th ARRG for 1892* (dated 29 March, 1894), PP 1893–94 XXI [c.7238]
1893 *56th ARRG for 1893* (dated 5 December, 1894), PP 1894 XXV [c.7551]
1894 *57th ARRG for 1894* (dated 5 December, 1895), PP 1895 XXIII Pt II [c.7768]
1895 *58th ARRG for 1895* (dated 10 December, 1896), PP 1897 XXI [c.8403]
1896 *59th ARRG for 1896* (dated 10 December, 1897), PP 1897 XXI [c.8591]
1897 *60th ARRG for 1897* (dated 10 December, 1898), PP 1898 XVIII [c.9016]
1898 *61st ARRG for 1898* (dated 9 December, 1899), PP 1899 XVI [c.9417]
1899 *62nd ARRG for 1899* (dated 31 December, 1900), PP 1900 XV [Cd 323]
1900 *63rd ARRG for 1900* (dated 1 March, 1902), PP 1901 XV [Cd 761]
1901 *64th ARRG for 1901* (dated 13 March, 1903), PP1902 XVIII [Cd 1230]
1902 *65th ARRG for 1902* (dated 1 July, 1904), PP 1904 XIV [Cd 2003]
1903 *66th ARRG for 1903* (dated 31 December, 1904), PP 1904 XIV [Cd 2197]
1904 *67th ARRG for 1904* (dated March 1906), PP 1905 XVII [Cd 2617]
1905 *68th ARRG for 1905* (dated December 1906), PP 1906 XX [Cd 3279]
1906 *69th ARRG for 1906* (dated December 1907), PP 1908 XVII [Cd 3833]
1907 *70th ARRG for 1907* (dated December 1908), PP 1909 X [Cd 4464]
1908 *71st ARRG for 1908* (dated December 1909), PP 1909 XI [Cd 4961]
1909 *72nd ARRG for1909* (dated January 1911), PP 1911 X [Cd 5485]
1910 *73rd ARRG for1910* (dated April 1912), PP 1911 XI [Cd 5988]
1911 *74th ARRG for 1911* (dated August 1913), PP 1912–13 XIII [Cd 6578]
1912 *75th ARRG for 1912* (dated Apr. 1914), PP 1913 XVII [Cd 7028]
1913 *76th ARRG for 1913* (dated March 1915), PP1914–16 IX [Cd 7780]
1914 *77th ARRG for 1914* (dated March 1916), PP 1916 V [Cd 8206]

1915 *78th ARRG for 1915* (dated March, 1917), PP 1917–18 V [Cd 8484]
1916 *79th ARRG for 1916* (dated December 1917), PP 1917–18 VI [Cd 8869]
1917 *80th ARRG for 1917* (dated February 1919), PP 1919 X [Cmd 40]
1918 *81st ARRG for 1918* (dated February 1920), PP 1920 X [Cmd 608]
1919 *82nd ARRG for 1919* (dated November 1920), PP 1920 XI [Cmd 1017]
1920 83rd ARRG 1920 (London, 1922 – undated)
This was the first and only ARRG that was not a Parliamentary Paper.

Registrar General's statistical reviews

1921 *Registrar General's statistical review (hereafter RGSR) for 1921* (London, 1923 – dated June 1923)
1922 *RGSR for 1922* (London, 1925 – dated September 1923)
1923 *RGSR for 1923* (London, 1925 – dated August 1924)
1924 *RGSR for 1924* (London, 1926 – dated October 1925)
1925 *RGSR for 1925* (London, 1927 – dated September1926)
1926 *RGSR for 1926* (London, 1928 – undated)
1927 *RGSR for 1927* (London, 1929 – undated)
1928 *RGSR for 1928* (London, 1930 – undated)
1929 *RGSR for 1929* (London, 1931 – undated)
1930 *RGSR for 1930* (London, 1932 – undated)
1931 *RGSR for 1931* (London, 1934 – undated)
1932 *RGSR for 1932* (London, 1935 – undated)
1933 *RGSR for 1933* (London, 1935 – undated); *Medical tables* (London, 1934); *Civil tables* (London, 1935)
1934 *RGSR for 1934* (London, 1936 – undated); *Medical tables* (London, 1935); *Civil tables* (London, 1936)
1935 *RGSR for 1935* (London, 1938 – undated); *Medical tables* (London, 1936); *Civil tables* (London, 1937)
1936 *RGSR for 1936* (London, 1938 – undated); *Medical tables* (London, 1937); *Civil tables* (London, 1938)
1937 *RGSR for 1937* (London, 1940 – undated); *Medical tables* (London, 1938); *Civil tables* (London, 1938)
1938 *RGSR for 1938 and 1939* (London, 1947 – undated); *Medical tables* (London, 1940); *Civil tables* (London, 1944)
1939 *RGSR for 1938 and 1939* (London, 1947 – undated); *Medical tables* (London, 1944); *Civil tables* (London, 1944)
1940 *RGSR for 1940–1945: Medical* (London, 1949 – undated); *RGSR for 1940–1945: Civil* (London, 1951 – undated); *Medical tables* (London, 1944); *Civil tables* (London, 1944)
1941 *RGSR for 1940–1945: Medical* (London, 1949 – undated); *RGSR for 1940–1945: Civil* (London, 1951 – undated); *Medical tables* (London, 1945); *Civil tables* (London, 1946)
1942 *RGSR for 1940–1945: Medical* (London, 1949 – undated); *RGSR for*

Appendix 1

1940–1945: Civil (London, 1951 – undated); *Medical tables* (London, 1946); *Civil tables* (London, 1947)

1943 *RGSR for 1940–1945: Medical* (London, 1949 – undated); *RGSR for 1940–1945: Civil* (London, 1951 – undated); *Medical tables* (London, 1948); *Civil tables* (London, 1948)

1944 *RGSR for 1940–1945: Medical* (London, 1949 – undated); *RGSR for 1940–1945: Civil* (London, 1951 – undated); *Medical tables* (London, 1949); *Civil tables* (London, 1949)

1945 *RGSR for 1940–1945: Medical* (London, 1949 – undated); *RGSR for 1940–1945: Civil* (London, 1951 – undated); *Medical tables* (London, 1947); *Civil tables* (London, 1948)

1946 *RGSR for 1946–1947: Medical* (London, 1949 – undated); *Medical tables* (London, 1948); *Civil tables* (London, 1949)

1947 *RGSR for 1946–1947 Medical* (London, 1949 – undated); *Medical tables* (London, 1949); *Civil tables* (London, 1949)

1948 *RGSR for 1948–1949: Medical* (London, 1953 – undated); *Medical tables* (London, 1950); *Civil tables* (London, 1950)

1949 *RGSR for 1948–1949: Medical* (London, 1953 – undated); *Medical tables* (London, 1951); *Civil tables* (London, 1951)

1950 *RGSR for 1950: Medical* (London, 1954 – undated); *Medical tables* (London, 1952); *Civil tables* (London, 1952)

1951 *RGSR for 1951: Text* (London, 1954); *Medical tables* (London, 1953); *Civil tables* (London, 1953)

Decennial supplements to the annual reports

1851–60
Supplement to the 25th annual report of the RG: Letter to the RG on the mortality in the registration districts of England during the 10 years 1851–60, by Wm Farr, PP 1865 XIII [3542]

1861–70
Supplement to the 35th annual report of the RG: Letter to the RG on the mortality in the registration districts of England during the 10 years 1861–70, by Wm Farr, PP 1875 XVIII [c. 1155–I]

1871–80
Supplement to the 45th annual report of the RG: Letter to the RG on the mortality in the registration districts of England during the 10 years 1871–80, by Wm Ogle, PP1884–85 XVII [c.4564]

1881–90
Supplement to the 55th annual report of the RG: Letter to the RG on the mortality in the registration districts of England during the 10 years 1881–90, by John Tatham. Pt I, PP 1895 XXIII [c.7769]; Pt II, PP 1897 XXI [c.8503]

Supplement to the 55th annual report of the RG: Letter to the RG on the mortality of males engaged in certain occupations in the three years 1890–92 and on an English healthy district life table for the ten years 1881–90, by John Tatham (London, 1897)

1891–1900

Supplement to the 65th annual report of the RG: Letter to the RG on the mortality in the registration districts of England and Wales during the 10 years 1891–1900, by John Tatham. Pt I, PP 1905 XVIII [Cd 2618]; Pt II, PP 1905 [Cd 2619]

1901–10

Supplement to the 75th annual report of the RG: Part I: Life Tables, PP 1914 XIV [Cd 7512]; *Supplement to the 75th annual report of the RG: Part II: Abridged Life Tables*, PP 1920, X [Cmd 1010]; *Supplement to the 75th annual report of the RG: Part III: Registration Summary Tables*, PP 1914–16 VIII [Cd 8002]; *Supplement to the 75th annual report of the RG: Part IV Mortality of Men in Certain Occupations in the Three Years 1910, 1911, and 1912* (London, n.d.)

1911–20

The Registrar General's Decennial Supplement, England and Wales 1921. Part I. Life Tables (London, 1927); *The Registrar General's Decennial Supplement, England and Wales 1921. Part II. Occupational Mortality, Fertility and Infant Mortality* (London, 1927); *The Registrar General's Decennial Supplement, England and Wales 1921. Part III. Estimates of population, statistics of marriages, births and deaths 1911–1920* (London, 1933)

1921–30

The Registrar General's Decennial Supplement, England and Wales 1931. Part IIa. Occupational Mortality (London, 1938); *The Registrar General's Decennial Supplement, England and Wales 1931. Part IIb. Occupational Fertility 1931 & 1939* (London, 1953); *The Registrar General's Decennial Supplement, England and Wales 1931. Part III. Estimates of population, statistics of marriage, births and deaths 1921–30* (London, 1952); *The Registrar General's Decennial Supplement, England and Wales 1931. Part IV. Multiple or secondary causes of death* (London, 1947)

1941–50

The Registrar General's Decennial Supplement, England and Wales 1951. Occupational Mortality. Part I (London, 1954); *The Registrar General's Decennial Supplement, England and Wales 1951. Occupational Mortality. Part II. Volume 1 Commentary* (London, 1958); *The Registrar General's Decennial Supplement, England and Wales 1951. Occupational Mortality. Part II. Volume 2 Tables* (London, 1958); *The Registrar General's Decennial Supplement, England and Wales 1951. Life tables* (London, 1957); *The Registrar General's Decennial Supplement, England and Wales 1951. Area Mortality* (London 1958)

Consolidated Bibliography

Manuscript sources

PRO: ACT 1: Government Actuary's Department: Correspondence and Papers.
PRO: CAB 134: Cabinet Office: Miscellaneous Committees: Minutes and Papers (General Series)
PRO: FD 1: Medical Research Committee and Medical Research Council: Files.
PRO: FD 2: Medical Research Committee and Medical Research Council: Annual Reports.
PRO: FD 4: Medical Research Committee and Medical Research Council: Reports of Special Research Projects.
PRO: FD 5: Medical Research Committee and Medical Research Council: Registered Files, Policy and Personnel Matters (PF Series) and Related Records.
PRO: HO 34: Public Offices Entry Books.
PRO: HO 44: Home Office: Correspondence, George IV and later.
PRO: HO 45: Home Office: Registered Papers.
PRO: HO 73: Home Office: Commissions, Various.
PRO: MH 13: General Board of Health: Correspondence.
PRO: MH 19: Poor Law Commission, Poor Law Board and Local Government Board: Government Offices Correspondence and Papers.
PRO: MH 78: Ministry of Health: Establishment and Organisation Files.
PRO: RG 19: GRO: Census Returns: Correspondence and Papers.
PRO: RG 20: GRO and Office of Population Censuses and Surveys, Establishment and Accounts Division: Correspondence and Papers.
PRO: RG 21: GRO: Local Registration Services: Correspondence and Papers.
PRO: RG 24: GRO: Royal Commission on Population, 1944–49.

PRO: RG 26: GRO: Population and Medical Statistics: Correspondence and Papers.
PRO: RG 27: GRO: Census Returns: Specimens of Forms and Documents.
PRO: RG 28: GRO: National Registration: Correspondence and Papers
PRO: RG 29 GRO: Letter Books.
PRO: RG 48: GRO: Registration of Births, Deaths and Marriages: Correspondence and Papers.
PRO: RG 50: GRO: Registrar General: Private Office Papers.
PRO: STAT 3: Stationery Office Out Letters.
PRO: STAT 12: Stationery Office: Files of Correspondence.
PRO: T 1: Treasury Board Papers.
PRO: T 162: Treasury: Establishments Department: Registered Files (E Series).
PRO: T 243: Treasury: Circulars and Minutes.
Scottish Record Office: GRO 5: Registration Branch files.
University College London Library: Papers of Sir Edwin Chadwick (1800–1890).

Parliamentary Papers

Hansard, 3rd Series.
Report of the Select Committee on Friendly Societies, PP 1825 IV.
Second report of the Select Committee on Friendly Societies, PP 1826–7 III.
Report of the Select Committee on Life Annuities, PP 1829 III.
First report of the Commissioners appointed to inquire into the law of England respecting real property , PP 1829 X.
Second report of the Commissioners appointed to inquire into the law of England respecting real property , PP 1830 XI.
Third report of the Commissioners appointed to inquire into the law of England respecting real property , PP 1831–2 XXIII.
Report of the select committee . . . on the general state of parochial registries and the laws relating to them, and on a general registration of births, baptisms, marriages, deaths and burials in England and Wales, PP 1833 XIV.
Fourth report of the Commissioners appointed to inquire into the law of England respecting real property , PP 1833 XXII.

Report of the commissioners . . . into the state of non-parochial registers, PP 1837–8 XXVIII.
1841 census report: abstract of the answers and returns, PP 1844 XXVII.
Eleventh Annual Report of the Registrar General for 1848, PP 1850 XX.
Copy of a report and tables, under the directions of the Lords of the Treasury, by the actuary of the National Debt Office, on the subject of sickness and mortality among the members of friendly societies, as shown by the quinquennial returns to 31 Dec. 1850, received by the Registrar of Friendly Societies in England, under the provisions of the Act 9 &10 Vict. c 27, PP 1852–53 C.
Copies of a further report and tables, under the directions of the Lords of the Treasury, by the actuary of the National Debt Office, on the subject of sickness and mortality among the members of friendly societies, as shown by the quinquennial returns to 31 Dec. 1850, received by the Registrar of Friendly Societies in England, under the provisions of the Act 9 &10 Vict. c 27, PP 1854 LXIII.
First Annual Report of the Registrar of Friendly Societies, PP 1856 LVIII.
First Report of the Medical Officer of the Privy Council, PP 1859 XII.
Report of the Select Committee on Civil Service Appointments, PP 1860 IX.
Return of the cost of reports and papers presented by command of Her Majesty . . . during the session of 1860, PP 1861 XXXV.
Fourth Report of the Medical Officer of the Privy Council, PP 1862 XXII.
1861 Census report, PP 1863 LIII, Pt 1.
11th Report of the Medical Officer of the Privy Council, PP 1868–9 XXXII.
First Report of the Royal Sanitary Commission, PP 1868–9 XXXII.
Second report of the Royal Sanitary Commission, Vol. 1, PP 1871 XXXV.
First report of the Civil Service Inquiry (Playfair) Commission, PP 1875 XXIII.
Census of England and Wales 1881. Vol. IV. General report, PP 1883 LXXX.

Conditions of the working classes. Tabulation of the statements made by men living in certain selected districts of London in March 1887, PP 1887 LXXI.
Report of the Treasury Committee on the Census, PP 1890, LVIII.
First and second report of the select committee on death certification, PP 1893–4 XI.
Report of the Departmental Committee on Old Age Pensions, PP 1898 XLV.
Report of the Interdepartmental Committee on Physical Deterioration, PP 1904 XXXII.
1901 census report, PP 1904 CVIII.
Census 1911, Vol. XIII, Fertility of marriage report, Part 1, PP 1917–18 XXXV.
Census 1911, Vol. XIII, Fertility of marriage report, Part 2 (HMSO, 1923).

Books and Articles

Abrams, Philip, *The origins of British sociology 1834–1914* (London, 1968).
Anderson, Margo J., *The American census. A social history* (London, 1988).
Anon., 'The tabulator. No 1', *The Engineer,* January 17, 1911, p. 96.
Anon., 'The tabulator. No 2', *The Engineer,* March 17, 1911, pp. 279–80.
Armstrong, David, 'The invention of infant mortality', *Sociology of Health and Illness,* 8 (1986), pp. 211–32.
Austoker, Joan, 'Walter Morley Fletcher and the origins of basic biomedical research policy', in Joan Austoker and Linda Bryder (eds), *Historical perspectives on the role of the MRC* (Oxford, 1989), pp. 23–34.
Austoker, Joan, and Bryder, Linda, 'The National Institute for Medical Research and related activities of the MRC', in Joan Austoker and Linda Bryder (eds.), *Historical perspectives on the role of the MRC* (Oxford, 1989), pp. 35–58.
Beaud, Jeanne-Pierre, and Prévost, Jean-Guy, 'La forme est le fond. La structuration des appareils statistiques nationaux (1800–1945)', *Revue de Synthèse,* 118 (1997), pp. 419–56.

Bellamy, Christine, *Administering central-local relations, 1871–1919. The Local Government Board in its fiscal and cultural context* (Manchester, 1988).
Beniger, James R., *The control revolution. Technology and the economic origins of the information society* (London, 1986).
Bentham, Jeremy, *Pauper management improved* (London, 1798).
Bentham, Jeremy, 'Outline plan of a general register of real property', in *The works of Jeremy Bentham published under the superintendence of his executor John Bowring* (London, 1859), vol. 5.
Bikson, Tora K., 'Organisational trends and electronic media: work in progress', *American Archivist*, 57 (Winter 1994), pp. 57–67.
Booth, Charles, *Life and labour of the people,* Series 1, 2 vols. (London, 1889).
Booth, Charles, *Life and labour of the people,* Series 2, 10 vols. (London, 1892–7).
Booth, Charles, *Life and labour of the people,* Series 3, 17 vols. (London, 1902).
Bowley, A. L., 'The importance of scientific method in statistical research', in R L Smyth (ed.) *Essays in economic method* (London, 1962), pp. 200–22.
Bowley, Marian *Housing and the State 1919–1944* (London, 1985).
Brand, Jeanne L., *Doctors and the State. The British medical profession and government action in public health, 1870–1912* (Baltimore, 1965).
Briggs, Asa, *Social thought and social action: a study of the work of Seebohm Rowntree 1871–1954* (London, 1961).
Brundage, Anthony, *England's "Prussian Minister". Edwin Chadwick and the Politics of Government Growth, 1832–1854* (London, 1988).
Bryder, Linda, 'Public health research and the MRC', in Joan Austoker and Linda Bryder (eds), *Historical perspectives on the role of the MRC* (Oxford, 1989), pp. 59–82.
Burn, John Southerden, The *Marriage and Registration Acts with instructions, forms, and practical directions for the use of officiating ministers, superintendent registrars, registrars, etc.* , (London, 1836).

Burns, John, 'Presidential address delivered to the First National Conference on Infantile Mortality on June 13, 1906', in G F McCleary, *The early history of the infant welfare movement* (London, 1933), pp. 151–68.

Campbell-Kelly, Martin, *ICL. A business and technical history* (Oxford, 1989).

Campbell-Kelly, Martin, 'Large-scale data processing in the Prudential, 1850–1930', *Accounting, Business and Financial History,* 2 (1992), pp. 117–39.

Campbell-Kelly, Martin, 'Charles Babbage and the assurance of lives', *IEEE Annals of the History of Computing,* 16 (1994), pp. 5–14.

Chadwick, Edwin, *An essay on the means of insurance against the casualties of sickness, decrepitude, and mortality: comprising an article reprinted from the Westminster Review (No. XVIII) from April 1828, with additional notes and comments* (London, 1836).

Chadwick, Edwin, 'On the best modes of representing accurately, by statistical returns, the duration of life, and the pressure and progress of the causes of mortality amongst different classes of the community, and amongst the population of different districts and countries', *Journal of the Statistical Society of London,* 7 (1844).

Clark, G. Kitson, "Statesmen in disguise': reflexions on the history of the neutrality of the civil service,' *The Historical Journal,* II, (1959), pp. 19–39.

Cobbett, William, *Rural rides* (Harmondsworth, 1983).

Cohen, Emmeline W., *The growth of the British Civil Service, 1780–1939* (London, 1965).

Coleman, B. I., 'The incidence of education in mid-century', in E A Wrigley (ed.) *Nineteenth-century society* (Cambridge, 1972), pp. 397–410.

Collini, Stefan, *Liberalism and sociology* (Cambridge, 1979).

Concise Dictionary of National Biography (Oxford, 1992).

Condorcet, Jean-Antoine-Nicolas de Caritat, marquis de, *Sketch for a historical picture of the progress of the human mind* (London, 1955).

Conk, Margo, 'Labor statistics in the American and English census: making some invidious comparisons', *Journal of Social History,* 16 (1983), pp. 83–102.

Cortada, James W., *Before the computer. IBM, NCR, Borroughs, and Remington Rand and the industry they created, 1865–1956* (Princeton, 1993).
Cortada, James W., 'Economic preconditions that made possible application of commercial computing in the United State', *IEEE Annals of the History of Computing*, 19 (1997), pp. 27–39.
Cowherd, Raymond G., *Political economists and the English Poor Laws. A historical study of the influence of classical economics on the formation of social welfare policy* (Athens, 1977).
Crowther, M. A. and White, Brenda M., 'Medicine, property and the law in Britain 1800–1914', *The Historical Journal*, 31 (1988), pp. 853–70.
Cullen, M. J., 'The making of the Civil Registration Act of 1836', *Journal of Ecclesiastical History*, XXV(1974), pp. 39–59.
Cullen, Michael J. *The statistical movement in early Victorian Britain* (Hassocks, 1975).
Curtis, Bruce, *The politics of population. State formation, statistics, and the census of Canada, 1840–1875* (London, 2001).
Daunton, M. J., *Progress and poverty. An economic and social history of Britain 1700–1850* (Oxford, 1995).
Davidson, R., and Lowe, R., 'Bureaucracy and innovation in British welfare policy 1870–1945', in W. J. Mommsen (ed.), *The emergence of the Welfare State in Britain and Germany* (London, 1981), pp. 263–95.
Davidson, Roger, *Whitehall and the labour problem in late-Victorian and Edwardian Britain* (London, 1985).
Davidson, Roger, 'Social intelligence and the origins of the welfare state', in Roger Davidson and Phil White (eds), *Information and government. Studies in the dynamics of policy making* (Edinburgh, 1988), pp. 14–37.
Desrosieres, Alain, 'Official statistics and medicine in nineteenth-century France: the SGF as a case study', *Social History of Medicine*, 4 (1991), pp. 515–38.
Desrosières, Alain, *La politique des grands nombres: histoire de la raison statistique* (Paris, 1993).
Dickens, Charles, *Bleak House* (London, 1852–3).
Dictionary of national biography (London, 1889).

Doll, R., and Hill, A. Bradford, 'Smoking and carcinoma of the lung', *British Medical Journal,* 30 September, 1950, pp. 739–48.

Doll, R., and Hill, A. Bradford, 'A study of the aetiology of carcinoma of the lung', *British Medical Journal,* 13 December, 1952, pp. 271–86.

Doll, R., and Hill, A. Bradford, 'Lung cancer and other causes of death in relation to smoking', *British Medical Journal,* 31 November, 1956, pp. 1071–81.

Donajgrodzki, A. P., "Social police' and the bureaucratic elite: a vision of order in the age of reform', in A P Donajgrodzki (ed.), *Social control in nineteenth century Britain* (London, 1977), pp. 51–76.

Duffield, Reginald, 'History of the Society of Medical Officers of Health', *Public Health* (Jubilee Number, 1906), pp. 1–5.

Dunn C. L., (ed.), *The Emergency Medical Services,* 2 vols (London, 1952).

Dupree, Marguerite W., 'Other than healing: medical practitioners and the business of life assurance during the nineteenth and early twentieth centuries', *Social History of Medicine,* 10 (1997), pp. 79–103.

Durbach, Nadja, "They might as well brand us': working-class resistence to compulsory vaccination in Victorian England', *Social History of Medicine* 13 (2000), pp. 45–62.

Dwork, Deborah, *War is good for babies and other young children. A history of the Infant and Child Welfare Movement in England 1898–1918* (London, 1987).

Eastwood, David '"Amplifying the province of the Legislature". The flow of information and the English State in the early nineteenth century', *Historical research,* LXXI (1989), pp. 276–94.

Eldridge, J. E. T., (ed.), *Max Weber: the interpretation of social reality* (London, 1970).

Elton, G. R., *Policy and police. The enforcement of the Reformation in the age of Thomas Cromwell* (Cambridge, 1972).

Emery, George, *Facts of life. The social construction of vital statistics, Ontario 1869–1952* (London, 1993).

Ende, Jan van den, 'The number factory: punched-card machines at the Dutch Central Bureau of Statistics', *IEEE*

Annals of the History of Computing, 16(3) (Fall, 1994), pp. 15–24.

Eyler, John M., 'Mortality, statistics and Victorian health policy: program and criticism', *Bulletin of the History of Medicine*, 50 (1976), pp. 335–55.

Eyler, John M., *Victorian social medicine. The ideas and methods of William Farr* (London, 1979).

Eyler, John M., 'The sick poor and the state: Arthur Newsholme on poverty, disease and responsibility', in Dorothy Porter and Roy Porter (eds) *Doctors, politics and society: historical essays – Clio Medica 23* (Amsterdam-Atlanta, 1993), pp. 188–211.

Eyler, John M., *Sir Arthur Newsholme and state medicine 1885–1935* (Cambridge, 1997).

Finer, S. E., *The life and times of Sir Edwin Chadwick* (London, 1952).

Finlayson, Geoffrey, *Citizens, state, and social welfare in Britain 1830–1990* (Oxford, 1994).

Flinn, M. W., 'Introduction', to Edwin Chadwick, *Report on the sanitary condition of the labouring population of Great Britain* (Edinburgh, 1965).

Freeden, Michael, 'Eugenics and progressive thought: a study in ideological affinity', *Historical Journal*, 22 (1979), pp. 645–54.

Freeden, Michael, *The New Liberalism. An ideology of social reform* (Oxford, 1978).

French, David, *British strategy and war aims 1914–1916* (London, 1986).

Fried, Albert, and Elman, Richard M., (eds), *Charles Booth's London. A portrait of the poor at the turn of the century, drawn from his 'Life and labour of the people in London'* (London, 1969).

Gilbert, Bently B., *The evolution of National Insurance in Great Britain: the origins of the Welfare State* (London, 1966).

Glass, D. V., 'A note on the occupational grouping used in tabulating the 1939 births', *The Registrar General's Decennial Supplement, England and Wales 1931. Part IIb. Occupational Fertility 1931 & 1939* (London, 1953), Appendix 7.

Glass, D. V., *Numbering the people. The eighteenth-century population controversy and the development of census and vital statistics in Britain* (London, 1973).

Goldman, Lawrence, 'Statistics and the society of science in early Victorian Britain; an intellectual context for the General Register Office', *Social History of Medicine,* 4 (1991), pp. 415–34.

Grieves, Keith, *The politics of manpower, 1914–19* (Manchester, 1988).

Greene, C. A. Everard, *The Beginnings. Reminiscences of C A Everard Greene* (British Tabulating Machine Company, London, c1959).

Greenhow, E. H., *Papers relating to the sanitary state of the people of England* (London, 1858; reprinted Farnborough, 1973).

Greenwood, Major, *The medical dictator and other biographical studies* (London, 1936).

Greenwood, Major, 'The occupational and economic factors of mortality', *British Medical Journal* (1939), pp. 862–6.

GRO, *Annual reports of the Registrar General:* see Appendix A.

GRO, *Supplements to annual reports of the Registrar General:* see Appendix A.

GRO, *Registrar General's statistical reviews:* see Appendix A.

GRO, *English life tables: tables of lifetimes, annuities, and premiums, with an introduction by William Farr* (London, 1864).

Gosden, P. H. J. H., *The friendly societies in England 1815–75* (Manchester, 1961).

Government Statisticians Collective, 'How official statistics are produced: views from the inside', in John Irvine, Ian Miles and Jeff Evans (eds), *Demystifying social statistics* (London, 1979), pp. 130–51.

Guinn, Paul, *British strategy and politics 1914 to 1918* (Oxford, 1965).

Hamlin, Christopher, 'Politics and germ theories in Victorian Britain: the Metropolitan Water Commissions of 1867–9 and 1892–3', in Roy MacLeod (ed.), *Government and expertise. Specialists, administrators and professionals, 1860–1919* (Cambridge, 1988), pp. 110–27.

Hannah, Leslie, *The rise of the corporate economy* (London, 1976).

Hannah, Leslie, and Kay, J. A., *Concentration in modern industry. Theory, measurement and the UK experience* (London, 1977).

Hannah, Matthew, *Governmentality and the mastery of territory in nineteenth century America* (Cambridge, 2000).

Hardy, Anne, 'Public health and the expert: the London Medical Officers of Health, 1856–1900', in Roy MacLeod (ed.), *Government and expertise. Specialists, administrators and professionals, 1860–1919* (Cambridge, 1988), pp. 128–44.

Hardy, Anne, *The epidemic streets. Infectious diseases and the use of preventative medicine 1856–1900* (Oxford, 1993).

Hardy, Anne, 'Death is the end of all disease: using the GRO cause of death statistics for 1837–1920,' *Social History of Medicine*, 7 (1994), pp 472–92.

Harrison, Ross, *Bentham* (London, 1983).

Heide, Lars, 'Shaping technology: American punched card systems 1880–1914', *IEEE Annals of the History of Computing*, 19(4) (Oct.–Dec. 1997), pp. 28–41.

Higgs, Edward, *Making sense of the census* (London, 1989).

Higgs, Edward, 'Diseases, febrile poisons, and statistics: the census as a medical survey', *Social History of Medicine*, 4 (1991), pp 465–78.

Higgs, Edward, *A clearer sense of the census. The Victorian censuses and historical research* (London, 1996).

Higgs, Edward, 'A cuckoo in the nest?: The origins of civil registration and state medical statistics in England and Wales' *Continuity and Change*, 11 (1996), pp. 115–34.

Higgs, Edward, 'The statistical Big Bang of 1911: ideology, technological innovation and the production of medical statistics', *Social History of Medicine*, 9 (1996), pp. 409–26.

Higgs, Edward, 'Medical statistics, patronage and the state: the development of the MRC Statistical Unit, 1911–1948', *Medical History* 44 (2000), pp. 323–40.

Higgs, Edward, 'The Annual Report of the Registrar General, 1839–1920: a textual history', in Magnello, Eileen and Hardy, Anne (eds.) *The Road to Medical Statistics* (Amsterdam and Atlanta, 2002), pp. 55–76.

Higgs, Edward, 'The General Register Office and the tabulation of data, 1837–1939', in Martin Campbell-Kelly, Mary Croarkin, John Fauvel and Raymond Flood (eds.), *From Sumer to spreadsheets: the curious history of tables* (Oxford, 2003), pp 209–34.

Higgs, Edward, *The Information State in England: the central collection of information on citizens since 1500* (Basingstoke, 2004).

Higgs, Edward, 'The linguistic construction of social and medical categories in the work of the English General Register Office', Simon Szreter, Arunachalam Dharmalingam and Hania Sholkamy (eds), *The qualitative dimension of quantitative demography* (Oxford, forthcoming)

Hill, Austin Bradford, *Principles of medical statistics* (London, 1937).

HMSO, *Chronological table of the Statutes covering the period from 1235 to the end of 1970* (London, 1971).

HMSO, *Imperial calendar for 1840* (London, 1840).

HMSO, *Imperial calendar for 1925* (London, 1925).

HMSO, *Imperial Calendar 1925–1934* (London, 1925–34).

Hobsbawm, E. J., 'The tramping artisan', *Economic History Review, 2nd Series,* III (1950–51), pp. 299–320.

Hogben, Lancelot, 'Major Greenwood', *Obituary notices of fellows of the Royal Society, Volume VII, 1950–51* (London, 1951), pp. 139–54;

Holdsworth, William, *A history of English law*, vol. 13 (London, 1952)

Hollerith, Herman, 'The electrical tabulating machine', *Journal of the Royal Statistical Society,* 57 (1894), pp. 678–82.

Honigsbaum, Frank, *The struggle for the Ministry of Health* (London, 1970).

Humphreys, Noel A., *Vital statistics: a memorial volume of selections from the reports and writings of William Farr* (London, 1885).

Humphreys, Noel A., 'Class mortality statistics', *Journal of the Royal Statistical Society* 50 (1887), pp. 255–92.

Hyman, Anthony, *Science and reform. Selected works of Charles Babbage* (Cambridge, 1989).

International Congress of Hygiene and Demography, *Transactions of the Seventh International Congress of Hygiene and Demography,* Vol. X, Div. II (London, 1892–3).

Jones, Gareth Stedman, *Outcast London. A study in the relationship between classes in Victorian Society* (Harmondsworth, 1984).

Jones, Greta, *Social Darwinism and English thought. The interaction between biological and social theory* (Brighton, 1980).

Jones, Greta, *Social hygiene in twentieth century Britain* (London, 1986).
Jordanova, Ludmilla, 'The social construction of medical knowledge'. *Social History of Medicine* 8 (1995), pp. 361–81.
Joyce, Patrick, *Visions of the people. Industrial England and the question of class 1848–1914* (Cambridge, 1991).
Joyce, Patrick, *The rule of freedom. Liberalism and the modern city* (London, 2003).
Ketelaar, Eric, 'Recordkeeping and office technology in Dutch public administration, 1823–1950', *Yearbook of European Administrative History* 9 (Baden-Baden, 1997), pp. 213–22.
La Berge, Ann, 'Edwin Chadwick and the French connection', *Bulletin of the History of Medicine*, 62 (1988), pp. 23–41.
Lambert, R. J., 'A Victorian National Health Service: state vaccination 1855–1871', *The Historical Journal*, V (1962), pp. 1–18.
Lambert, Royston, *Sir John Simon and English social administration 1816–1904* (London, 1963).
Larson, Edward J., 'The rhetoric of eugenics: expert authority and the Mental Deficiency Bill', *British Journal for the History of Science*, 24 (1991), pp. 45–60.
Lawton, Richard, 'Census data for urban areas', in Richard Lawton (ed.), *The census and social structure* (London, 1978), pp. 82–145.
Lewes, Fred, 'William Farr and the communication of cholera', *William Farr 1807–1883. Commemorative symposium* (OPCS Occasional Paper 33, 1985), p. 10.
Lewes, Fred, 'The GRO and the provinces in the nineteenth century,' *Social History of Medicine*, 4, (1991), pp 479–96.
Lewis, Jane, *The politics of motherhood. Child and maternal welfare in England, 1900–1939* (London, 1980)
Lewis, R. A., *Edwin Chadwick and the Public Health Movement 1832–1854* (London, 1952).
Lewis-Faning, E., 'A study of the trend of mortality rates in urban communities of England and Wales, with special reference to depressed areas', *British Medical Journal*, 24 April, 1937, pp. 865–7.
Lemmings, David, 'Marriage and the law in the eighteenth century: Hardwicke's Marriage Act of 1753', *The Historical Journal*, 39(1996), pp. 339–60.

Logan, W. P. D., and Brooke, Eileen M., *The survey of sickness, 1943 to 1952. General Register Office Studies on Medical and Population Subjects, no. 12* (London, 1952).

Lokke, Anne, 'No difference without a cause. Infant mortality rates as a world view generator', *Scandinavian Journal of History*, 20 (1995), pp. 75–96.

Luckin, Bill, 'Death and survival in the city: approaches to the history of disease', *Urban History Yearbook* (Leicester, 1980), pp. 53–62.

McBriar, A. M., *An Edwardian mixed doubles: the Bosanquets versus the Webbs. A study in British social policy 1890–1929* (Oxford, 1987).

McCulloch, John Ramsay, *A statistical account of the British Empire: exhibiting its extent, physical capacities, population, industry, and civil and religious institutions*, vol. 2 (London, 1837).

McKinlay, Peter L., 'Infant mortality and economic status', *Lancet*, 3 November 1928, pp. 938–40.

MacKenzie, Donald A., *Statistics in Britain, 1865–1930: the social construction of scientific knowledge* (Edinburgh, 1981).

MacLeod, Roy M., 'Law, medicine and public opinion: the resistance to compulsory health legislation 1870–1907. Part I and II', *Public Law* (1967), pp. 107–28, 189–211.

MacLeod, Roy M., 'The frustration of state medicine 1880–1899', *Medical History*, 11 (1967), pp. 15–40.

MacLeod, Roy M., *Treasury control and social administration. LSE Occasional Papers on Social Administration, no 23* (London, 1968).

MacLeod, Roy M., 'Introduction', in Roy MacLeod (ed.) *Government and expertise. Specialists, administrators and professionals, 1860–1919* (Cambridge, 1988), pp. 1–26.

Magnello, M. Eileen, 'Karl Pearson's Gresham lectures: W.F.R. Weldon, speciation and the origins of Pearsonian statistics', *British Journal of the History of Science*, 29 (1996), pp. 43–63.

Magnello, M. Eileen, 'The non-correlation of biometrics and eugenics: rival forms of laboratory work in Karl Pearson's career at University College London', *History of Science* 37 (1999), pp. 79–106, 123–50.

Mallet, Bernard, *British budgets, 1887–88 to 1912–13* (London: 1913).

Mallet, Bernard, 'The organisation of registration in its bearings on vital statistics', *Journal of the Royal Statistical Society,* LXXX (January, 1917), pp. 1–24.

Mallet, Bernard, 'Is England in danger of racial decline?', *National Review* (Feb. 1922), pp. 843–53.

Mallet, Bernard, 'Registration in relation to eugenics', *Eugenics Review,* 14 (1922), pp. 23–30.

Manchester, A. H., *Modern legal history* (London, 1980).

Matthews, J. Rosser, *Quantification and the quest for medical certainty* (Princeton, 1995).

Mazumdar, Pauline M. H., *Eugenics, human genetics and human failings. The Eugenics Society, its sources and its critics in Britain* (London, 1992).

Merriam, W. R., 'The evolution of American census-taking', *Century Magazine* (April 1903), pp. 840–1.

Mill, John Stuart, and Bentham, Jeremy, *Utilitarianism and other essays* (London, 1987).

Ministry of Health, *Annual report of the chief medical officer of the Ministry of Health 1919–20* (London, 1920).

Ministry of Health, *On the state of the Public Health. Annual report of the chief medical officer of the Ministry of Health for the year 1932* (London, 1933).

Ministry of Health, *On the state of the Public Health. Annual report of the chief medical officer of the Ministry of Health for the year 1933* (London, 1934).

Ministry of Health, *On the state of the Public Health. Annual report of the chief medical officer of the Ministry of Health for the year 1934* (London, 1935).

Mitchison, Rosalind, *British population change since 1860* (London, 1977).

Mooney, Graham, 'Still-births and the measurement of urban infant mortality rates c. 1890–1930', *Local Population Studies,* 53 (1994), pp. 42–52.

Mooney, Graham, 'Professionalization in public health and the measurement of sanitary progress in nineteenth-century England and Wales', *Social History of Medicine* 10 (April 1997), pp. 53–78.

Mooney, Graham, Luckin, Bill, and Tanner, Andrea, 'Patient

pathways: solving the problem of institutional mortality in London during the later nineteenth century', *Social History of Medicine* 12 (1999), pp. 227–70.

Moriyama, Iwao M., Loy, Ruth M., Robb-Smith, A. H. T., *History of nomenclature of diseases, the International Classification of Diseases, and the Classification of Causes of Death* (London, 1994).

Moss, Louis, *The Government Social Survey: a history* (London, 1991).

National Conference on Infant Mortality, *Report of the proceedings of the National Conference on Infantile Mortality ... 23rd, 24th, and 25th March 1908* (London, 1908).

Nissel, Muriel, *People Count. A History of the General Register Office* (London, 1987).

Office of Population Censuses and Surveys and the General Register Office, Edinburgh, *Guide to Census Reports, Great Britain 1801–1966* (London, 1977).

Ogle, William, 'Proposal for the establishment and international use of a standard population, with fixed sex and age distribution, in the calculation and comparison of marriage, birth and death rates', *Bulletin de l'Institute International de Statistique, Rome*, VI (1892), pp. 83–5.

Pearson, Karl, 'On the inheritance of the mental and moral characters in man, and its comparison with the inheritance of the physical characters. The Huxley Lecture for 1903', *Journal of the Anthropological Institute of Great Britain and Ireland*, 33 (1903), pp. 179–237.

Pearson, Karl, *Medical progress and eugenics* (London, 1912).

Pearson, Robin, 'Thrift or dissipation? The business of life assurance in the early nineteenth century,' *Economic History Review*, 2nd series, XLIII (1990), pp. 236–254.

Perrot Jean-Claude & Woolf, Stuart J., *State and statistics in France 1789–1815* (New York, 1984).

Porter, Dorothy, '"Enemies of the race": biologism, environmentalism, and public health in Edwardian England', *Victorian Studies*, XXXIV (1991), pp. 159–78.

Porter, Dorothy, 'Stratification and its discontents: professionalization and conflict in the British public health service', in

Elizabeth Fee and Roy M. Acheson (eds), *A history of education in public health. Health that mocks the doctors' rules* (Oxford, 1991), pp. 83–113.

Porter, Theodore M., *The rise of statistical thinking 1820–1900* (Princeton, 1986).

Postema, Gerald J., *Bentham and the common law tradition* (Oxford, 1986).

Riley, James C., 'Ill health during the English mortality decline: the friendly societies experience', *Bulletin of the History of Medicine*, 61 (1987), pp. 563–88.

Riley, James C., 'Disease without death: new sources for a history of sickness', *Journal of Interdisciplinary History*, 17 (1987), pp. 537–63.

Riley, James C., 'The morbidity of medical practitioners', *Social History of Medicine*, 9 (1996), pp. 467–71.

Riley, James C., *Sick, not dead: the health of British working men during the mortality decline* (Baltimore, 1997).

Rogers, Everett M., *The diffusion of innovations,* (London, 1983).

Roseveare, Henry, *The Treasury: the evolution of a British institution* (London, 1969).

Rowntree, B. Seebohm, *Poverty : a study of town life* (London, 1901).

Ruggles, Steven, Sobek, Matthew, and Gardner, Todd, 'Distributing large historical census samples on the Internet', *History and Computing,* 8 (1996), pp. 145–59.

Sampson, Anthony, *Company man. The rise and fall of corporate life* (London, 1995).

Sauer, R., 'Infanticide and abortion in nineteenth-century Britain', *Population Studies,* 32 (1978), pp. 81–93.

Schürer, Kevin 'The 1891 census and local population studies', *Local Population Studies* 47 (1991), pp. 16–29.

Searle, G. R., *The quest for national efficiency. A study in British Politics and political thought, 1899–1914* (London, 1971).

Searle, G. R., *Eugenics and politics in Britain 1990–1914* (Leyden, 1976).

Simon, John, *English sanitary institutions* (London, 1897).

Simpson, Helen, 'The Management of Electronic Information Resources in a Corporate Environment', in Edward Higgs (ed.), *History and electronic artefacts* (Oxford, 1998), pp. 87–100.

Snellen, Ignace Th. M., 'From societal scanning to policy feedback: two hundred years of government information processing in the Netherlands', *Yearbook of European Administrative History 9* (Baden-Baden, 1997), pp. 195–212.

Soloway, Richard A., *Demography and degeneration. Eugenics and the declining birthrate in twentieth-century Britain* (London, 1990).

Stacey, Stephen, 'The Ministry of Health 1919–1929: ideas and practice in a government department' (D Phil thesis, Oxford, 1984).

Stern, J., 'Social mobility and the interpretation of social class mortality differentials', *Journal of Social Policy*, 12 (1983), pp. 27–49.

Stevenson, T. H. C., 'Suggested lines of advance in English vital statistics', *Journal of the Royal Statistical Society*, LXXIII (1910), pp. 685–702.

Stevenson, T. H. C., 'The fertility of various social classes in England and Wales from the middle of the nineteenth century to 1911', *Journal of the Royal Statistical Society*, 83 (1920), pp. 401–44.

Stevenson, T. H. C., 'The social distribution of mortality from different causes in England and Wales, 1910–12', *Biometrika*, 15 (1923), pp. 382–400.

Stevenson, T. H, C., 'The vital statistics of wealth and poverty', *Journal of the Royal Statistical Society*, 91 (1928), pp. 401–44.

Stewart, Robert Mackenzie, *Henry Brougham, 1778–1868: his public career* (London, 1986).

Stocks, Percy, 'Fresh evidence on the inheritance factor in tuberculosis', *Annals of Eugenics*, 3 (1928), pp. 84–95.

Stocks, Percy, 'Infant mortality in the metropolitan boroughs in relation to occupation', *Annals of Eugenics*, 3 (1928), pp. 194–200.

Stocks, Percy, 'A study of the epidemiology of measles', *Annals of Eugenics*, 3 (1928), 361–98.

Stocks, Percy, 'A biometric investigation of twins and their brothers and sisters', *Annals of Eugenics*, 4 (1930–31), pp. 49–108.

Stocks, Percy, 'The distribution of cancer and tuberculosis mortality in England and Wales', *Annals of Eugenics*, 4 (1930–31), pp. 341–61.

Stocks, Percy, 'On the spread of small-pox in partially vaccinated communities', *Annals of Eugenics*, 5 (1932), pp. 302–10.

Supple, Barry, 'Legislation and virtue: an essay on working class self-help and the State in the early nineteenth century', in Neil McKendrick (ed.) *Historical perspectives. Studies in English thought and society in honour of J H Plumb* (London, 1974), pp. 215–254.

Szreter, Simon, 'The genesis of the registrar-general's social classification of occupations', *The British Journal of Sociology*, XXXV (1984), pp. 522–46.

Szreter, Simon, 'Introduction: the GRO and the historians', *Social History of Medicine*, 4 (1991), pp. 401–15.

Szreter, Simon, 'The GRO and the public health movement in Britain 1837–1914,' *Social History of Medicine*, 4 (1991), pp. 435–64.

Szreter, Simon, *Fertility, class and gender in Britain 1860–1940* (Cambridge: CUP, 1996).

Thane, Pat, *The foundations of the Welfare State* (London, 1982).

Thompson, David M., 'The religious census of 1851', in Richard Lawton (ed.), *The census and social structure* (London, 1978), pp. 241–86.

Thompson, F. M. L., *English landed society* (London, 1963).

Thomson, A. Landsborough, *Half a century of medical research. Volume 1: origins and policy of the Medical Research Council (UK)* (London, 1973).

Thomson, A. Landsborough, *Half a century of medical research. Volume 2: the programme of the Medical Research Council (UK)* (London, 1975).

Thomson, Matthew, 'Sterilization, segregation and community care: ideology and solutions to the problem of mental deficiency in inter-war Britain,' *History of Psychiatry*, 3 (1992), pp. 473–98.

Titmuss, Richard M., *Poverty and population. A factual study of contemporary social wastage* (London, 1938).

Titmuss, Richard M., *Birth, poverty and wealth* (London, 1943).

Turner, John, "Experts' and interests: David Lloyd George and the dilemmas of the expanding state, 1906–19', in Roy MacLeod (ed.), *Government and expertise. Specialists, administrators and professionals, 1860–1919* (Cambridge, 1988), pp. 203–23.

Waddams, S. M., *Law, politics and the Church of England; the career of Stephen Lushington, 1782–1873* (Cambridge, 1992).

Webster, Charles, 'Healthy or hungry thirties', *History Workshop Journal*, 13 (1982), pp. 110–29.

Whitehead, Frank, 'The Government Social Survey', in Martin Bulmer (ed.), *Essays in the history of British sociological research* (Cambridge, 1985), pp. 83–100.

Who was who, 1971–1980 (London, 1981).

Wilkinson, Richard G., 'Socio-economic differences in mortality: interpreting the data on their size and trends' in Richard G Wilkinson (ed.), *Class and health. Research and longitudinal data* (London, 1986), pp. 1–20.

Wilkinson, Richard G., *Unhealthy societies. The afflictions of inequality* (London, 1996).

Williams, Naomi, 'The implementation of compulsory health legislation: infant smallpox vaccination in England and Wales 1840–1890', *Journal of Historical Geography*, 20 (4) (Oct. 1994), 396–412.

Winter, J. M., *The Great War and the British People* (London, 1986).

Wohl, Anthony S., *Endangered lives: public health in Victorian Britain*(London, 1983).

Woods, Robert, 'Physician, heal thyself: the health and mortality of Victorian doctors', *Social History of Medicine*, 9 (1996), pp. 1–30.

Woods, Robert, "Sickness is a baffling matter'. A reply to James C Riley', *Social History of Medicine*, 10 (1997), pp. 157–63.

Wright, Maurice, *Treasury control of the Civil service 1854–1874* (Oxford, 1969).

Wrigley, E. A. and Schofield, R. S., *The population history of England 1541–1871* (London, 1981).

Yates, JoAnne, *Control through communication : the rise of system in American management* (Baltimore, 1989).

Yelling, J. A., *Slums and redevelopment. Policy and practice in England, 1918–45* (London, 1992).

Zimmeck, Meta, 'Strategies and stratagems for the employment of women in the British Civil-Service, 1919–1939', *Historical Journal*, 27 (1984), pp. 901–24.

Zuboff, Shoshana, *In the age of the smart machine: the future of work and power* (London, 1988).

Index

1753 Clandestine Marriages Act, 3
1819 Friendly Societies Act, 26
1829 Friendly Societies Act, 26
1836 Marriage Act, 1, 2, 14, 16
1836 Registration Act, 1–29, 59–60, 110, 171, 192, 218
1840 Non-Parochial Registers Act, 5
1840 Vaccination Act, 78, 85
1846 Friendly Societies Act, 39
1848 Public Health Act, 38, 57, 79
1853 Vaccination Extension Act, 85
1855 Metropolis Management Act, 84
1857 Extra-Parochial Places Act, 110
1858 Births and Deaths Registration Act, 5
1866 Exchequer and Audit Departments Act, 103
1867 Vaccination Act, 85
1872 Infant Life Protection Act, 86
1872 Public Health Act, 84
1874 Births and Deaths Registration Act, 19, 86, 88, 120
1876 Divided Parishes and Poor Law Amendment Act of 1876, 111
1879 Poor Law Act, 111
1882 Divided Parishes Act, 111
1888 Local Government Act, 84
1889 Infectious Diseases (Notification) Act, 20
1894 London Equalisation of Rates Act, 112
1898 Marriage Act, 151
1898 Vaccination Act, 85
1908 Old Age Pensions Act, 153
1911 National Insurance Act, 153, 167, 184
1915 National Registration Act, 187
1918 Maternal and Child Welfare Act, 136
1920 Census Act, 76, 160, 201
1929 Local Government Act, 50
1938 Population (Statistics) Act, 210

Abrams, Philip, 62
actuarial life tables, 6, 27–8, 34–40, 53, 60, 101, 183, 223
Addison, Sir Christopher, 189, 191–4, 196, 201
administrative boundaries, 41, 98, 110–12, 151, 171–3
Advisory Committee on Scientific Policy, 213
Annals of Eugenics, 181, 208
Annual Report of the Registrar General, 51–6, 90–1, 129–30, 221–6
Anti-Vaccination League, 85, 87

Babbage, Charles, 28
bacteriology, 93–5, 207

Bankruptcy and Insolvent
  Debtors' Courts, 74
Bedlam, 75
Belinge, Henry, 23
Bellingham, Archer, 162–3
Bentham, Jeremy, 9, 11, 18, 27–8
Biometric Laboratory, 179–80
Biometrika, 99, 178
Board of Customs and Excise,
  153
Board of Education, 77, 133,
  180, 189, 191
Board of Trade, 104, 127, 163,
  165, 193
Boer War, 98–9, 135
boot and shoe industry, 94, 185,
  203
Booth, Charles, 124–5, 139,
  153
Bowley, A. L., 65, 191
British Association for the
  Advancement of Science,
  57, 63–4, 79
British Medical Journal, 197
British Tabulating Machine
  Company, 167, 170
Brougham, Henry Peter, Baron
  Brougham and Vaux, 4, 7,
  9, 13–14
Brownlee, John, 94, 184–5,
  202–3, 206
Buchanan, George, 118, 175,
  195–6
Burn, John Southerden, 16
Burns, John, 101, 135

Cannan, Edwin, 135
Census, Treasury Committee on
  the (1890), 124–7
censuses,
  1841 census, 40–3

1911 fertility survey, 130–1,
  139, 144
1939 'census', 209
administrative boundaries
  and, 110–12
as a medical survey, 57
effects on GRO publications,
  51–3
industrial census, 163
machine tabulation and, 140,
  157–73
occupational classification in,
  36
permanent legislation for,
  200–1
poverty in London, 117, 143
role of Major George
  Graham in, 75–77
Treasury Committee on,
  124–7
Central Probate Registry, 10,
  17
Central Register of War
  Refugees, 187
Chadwick, Edwin, 22–32, 43,
  45, 58–9, 79, 81–3, 115,
  218
Chamberlain, Joseph, 118
Charity Organisation Society,
  141
Churchill, Winston, 201
Civil Service Appointments,
  Select Committee on
  (1860), 73–4
Civil Service Commission, 106
Civil Service, Playfair
  Commission of inquiry on
  the (1874), 107
Clarke, Sir James, 32
Cobbett, William, 26
Cortada, James W., 166

County Councils Association, 171
Cromwell, Thomas, 1, 10
Cullen, Michael J., 4–6, 12, 24–5

Darwin, Charles, 96
Davidson, Roger, 181, 199
Death Registration, Select Committee on (1893), 19
Decennial Supplements to the Annual Report of the Registrar General, 90, 227–8
Derrick, V. A. P., 196, 207
Dickens, Charles, 9
Dissenters. See Nonconformists
Dunbar, Sir William, 129, 137, 150–2
Durkheim, Emile, 92

Eicholz, Alfred, 133
Emergency Hospital Service, 211
Employment of Supplementary Clerks, Committee on the (1860), 105
Ende, Jan van den, 169
eugenics, 64, 92–101, 129–41, 144–5, 149–51, 156, 163, 175, 178–81, 208
Eugenics Record Office, 179
Eugenics Society, 95, 139,
Evolutionary Committee of the Royal Society, 100
Eyler, John, 6–7, 25, 39, 43, 56, 60, 66–7, 120,

Farr, William,
    actuarial work of, 34–9
    early career of, 32–3

George Graham and, 67–77
hospital mortality and, 173
Letter to the Registrar General, 53–6
morbidity and, 20, 27
post of Registrar General and, 119–20
smallpox vaccination and, 87, 116
statistical theory of, 56–64, 97, 142
stillbirths and, 19
fertility, 56, 96–100, 130–3, 138–41, 144–9, 157, 163–4, 186, 210–11, 228
Finer, S. E., 43, 152
Finlayson, Alexander, 40
Finlayson, John, 27, 34, 40
Fisher, R. A., 180
Fletcher, Sir Walter, 183, 195, 204–7
Flinn, Michael, 120
France, 17–18, 111, 160, 164, 168–9
Fraser, D. C., 183
friendly societies, 25–9, 34–5, 39–40

Galton, Sir Francis, 63–4, 95–6, 157, 178–81, 207–8
General Board of Health, 79–83, 115
General Post Office, 38, 59, 72
Germany, 97–8, 135, 153, 166, 187, 210
Gladstone, William, 103, 111, 153
Glass, D. V., 6, 23
Goldman, Lawrence, 5
Government Actuaries Department, 183, 206

Graham, George,
  actuarial project of, 34–8
  morbidity and, 20
  property rights and, 16–17
  opposition to Civil Service reform, 105–7
  role in General Register Office, 67–83
  stillbirths and, 19
Great Eastern Railway, 38
Greene, Everard, 167
Greenwood, Major, 70, 77, 145, 179–81, 195–6, 202–3, 206–7
Grimshaw, T. W., 134, 143

Haldane, J. B. S., 180
Hardy, Anne, 91
Hardy, G. F., 183
Hawkins, Francis Bisset, 23
Health of Towns and Populous Places, Royal Commission on the (1843), 38
Henniker, Sir Brydges,
  closeness to Local Government Board, 116–19
  eugenics and, 100–1
  weakness of, 107–8, 119–28
Hill, Roland, 59
Hogben, Lancelot, 180
Hollerith punch tabulators, 159–61, 164–70, 176, 178
Hollerith, Herman, 159–60, 168
Home Office, 19, 23, 25, 41, 60, 76, 80, 88, 113, 116, 163, 165
home working, 163
Hospital In-Patient Enquiry, 213
hospital mortality, 109, 131, 170–4, 184, 213

Humphreys, Noel, 32, 66, 118, 123, 134, 141–4

Incorporated Society of Medical Officers of Health, 84
Industrial Health Research Board, 204
infant mortality, 19–20, 85–7, 98, 130–9, 143–6, 180, 199
inspection of local registration, 75
Institute of Actuaries, 39, 182–3
International Congress of Hygiene and Demography, 142–3
International List of the Causes of Death, 21, 131, 213
International Statistical Congress, 61–2
Isserlis, Leon, 207

Jones, Greta, 138

King, George, 182
Koch, Robert, 93

Laboratory of National Eugenics, 140, 179–80, 207
Lamarck, Jean-Baptiste, 96
Lambert, John, 67, 82–5, 114
Lancet, 117, 181, 197
Land Registry, 10
Laycock, Dr John, 23, 32
League of Nations, 198
Letheby, Henry, 59
Lidstone, G. J., 183
Liebig, Justus von, 57, 93
Lister, Thomas,
  origins of statistical role of the General Register Office and, 29–34

1841 census and, 41–3
administrative weakness of, 72–3
Lloyd George, David, 191–2, 194
Local Government Board, absorption into Ministry of Health, 188–90
infant mortality and, 135–7
poverty survey in London and, 117, 143
staffing, 152
subordination of General Register Office to, 111–27
local registrars of births, marriages and deaths, 2, 24, 47, 49–50, 75–6, 85, 116, 171, 174, 187, 210
London School of Hygiene and Tropical Medicine, 94, 202–3, 206–7
Lowe, Robert, 69
Lushington, Stephen, 15

Machinery of Government (Haldane) Committee (1918), 191
MacLeod, Sir Reginald, 59, 85, 102, 114–15, 150–2
Mallet, Sir Bernard,
actuarial role and, 181–2
Annual Report of the Registrar General and, 130
eugenic interests of, 137–40, 149
machine tabulation and, 157–64
national registration and, 187–8
on origins of the General Register Office, 17
resignation of, 188–93
Marshall, Alfred, 124–5, 127
mass radiography programme, 214
Maurice, General Sir John Frederick, 98
McCulloch, John Ramsay, 32
Medical Department of the Privy Council, 59, 82–3, 113, 115, 170
medical officers of health, 49, 59, 82–4, 92, 101, 113, 126, 135–7, 140, 170, 174–5, 203
Medical Research Council, 93–4, 183–5, 191, 195, 200–8, 212
Merriam, W. R., 159, 168
Metropolitan Association of Medical Officers of Health, 84
Ministry of Health, 67, 149, 155, 161, 175, 180, 185–207, 212–14
Ministry of Labour, 197
Ministry of National Insurance, 212
Monro, Sir Horace, 189
Mooney, Graham, 19
Moore, Dr James, 13
Morant, Sir Robert, 189–96, 200–1
morbidity, 18, 20–1, 26–7, 39–40, 56, 142, 209–14, 218

National Conferences on Infant Mortality, 135
National Debt Office, 27, 39–40

National Expenditure, Select Committee on (1902), 104
National Health Insurance Commission, 189–90
National Health Insurance Joint Committee, 154, 183
National Health Service. *See* NHS
National Institute for Medical Research, 94, 184
national insurance, 153–4, 167, 183–4, 189
National Insurance Joint Committee, 184
national registration, 154, 186–8, 209–10
National Registration, Committee on (1917), 187–8
National Registration, Jackson Committee on (1915), 187
National Registration, Landsdowne Committee on (1915), 187
National Union of Boot and Show Operatives, 203
Netherlands, 164, 168–9
New Liberalism, 149–55
Newman, George, 180, 189–90, 194–6, 205–7
Newsholme, Arthur, 100–1, 115, 132, 139–40, 145, 150, 180, 189, 198
NHS, 210, 213–14
NHS Central Register, 210
Nightingale, Florence, 38, 57
Nonconformists, 3–6, 10–16, 77, 162
non-parochial registers, 5
Northcote-Trevelyan reforms of the Civil Service, 69

nosologies, 57, 65, 131, 176
notifiable diseases, 212
Nugent, George Greville, Baron Nugent of Carlanstown, 4, 7, 11

occupational classifications, 36, 124
Ogle, William, 19, 90–2, 101, 109, 118, 126, 141–3, 216, 227
old age pensions, 151–3

Parochial Registries, Select Committee on (1833), 4–7, 11–16, 23
parochial registration, 1–17
Patent Office, 10
Pearson, Karl, 63, 95–100, 132, 139–40, 157, 178–80, 184, 207–8
Penrose, Lionel, 180
Phipps, Edmund, 42
Physical Deterioration, Interdepartmental Committee on (1903), 98–100, 133
Playfair, Lyon, 86–7, 107
Poor Law Commissioners, 23–5
Population, Royal Commission on (1943), 211
Porter, Theodore, 61
poverty survey in London, 117–8, 143
Pratt, John Tidd, 39–40
Price, Richard, 35
property rights registration, 7–17, 45–50
Protection of Infant Life, Committee on the (1871), 86

Protestant Society for the
    Protection of Religious
    Liberty, 4
Provincial Medical and Surgical
    Association, 6, 23
Prudential Insurance Company,
    166–7
Public Record Office, 10, 19

Quetelet, Adolphe, 6, 13

Radium Commission, 213
Real Property, Commission of
    enquiry on (1829), 8,
    10–11, 14
Registry of Friendly Societies,
    39–40
Registrar General's Statistical
    Review, 199–201, 226–7
Rickman, John, 40–2
Robinson, Sir Arthur, 191,
    196–8
Rogers, Everett, 157–8, 160,
    165, 170
Rose, Sir George, 3, 26
Rowntree, Benjamin Seebohm,
    153
Royal College of Physicians, 31
Royal College of Surgeons, 31
Royal Insurance Co., 183
Royal Statistical Society, 17, 57,
    134, 145, 159–60, 168, 181
Rumsey, Henry, 59
Russell, Lord John, 1st Earl
    Russell, 5, 14, 18, 24, 28,
    41

Saleeby, Caleb, 96
Sanitary Commission, Royal
    (1871), 16, 19–20, 82, 86
Schiller, F. C. S., 99

Sclater-Booth, George, 86, 116
Scotland, 87, 168, 183, 187
separation allowances, 154
Simon, John, 59, 67, 81–4,
    113–15
smallpox vaccination, 78, 80–87,
    116
Smith, Thomas Southwood, 63
Snow, E. C., 181–2
social darwinism, 92, 95, 138–9
social hygiene, 138–9
Society of Antiquaries, 49
Society of Apothecaries, 31–2
socio-economic class, 130, 133,
    141–9
Somerset House, 17, 47–9, 118
Spencer, Herbert, 70, 96
staffing of the General Register
    Office, 46–50, 64–7, 70–5,
    102–9, 149–52, 176–8, 188
standardisation of mortality rates,
    91
State Munitions Establishments,
    155
Stationery Office, 52, 151
statistical limitations of the
    General Register Office,
    64–7, 178–85, 194–8,
    202–7
Statistical Society of London.
    See Royal Statistical
    Society
Statistique Gènèrale de la
    France, 17
Stevenson, T. H. C.,
    early career, 132
    frustration of later career, 176,
    195–8, 203,
    machine tabulation and,
    161–2
    national registration and, 187

socio-economic groupings and, 133, 138–41, 144–50
stillbirths, 18–19, 218
Stocks, Percy, 207–8, 211–17
Supreme Court of Judicature, 9
Survey of Sickness, 211–12
Szreter, Simon, 53, 56–7, 83, 91–5, 101, 119, 124–5, 127, 132–4, 139, 143–4

Tatham, John, 90–2, 101, 129, 135, 137, 143, 171, 216
tenancy by courtesy, 8
Titmuss, Richard, 148, 200
Topley, W. W. C., 203
Treasury,
    1841 census and, 42
    Census Committee (1890) and, 123–7
    distribution of General Register Office reports and, 88
    Edwardian expansion of Civil Service and, 150–4
    financial stringency of, 102–9
    opinion of Brydges Hennicker, 122–3
    machine tabulation and, 140, 160–5, 172
    Major George Graham's relationship with, 68–9, 72–4
    statistical role of the General Register Office and, 29–33
    permanent Census Act and, 200–1

United States of America, 97–8, 135, 157, 164, 166–8

Vaccination, Committee on (1871), 86
Vardon, Thomas, 42
Vivian, Sylvanus, 190–201, 209, 216

Walpole, Spencer, 70
war widow's pensions, 154
Watson, Sir Alfred, 183
Weismann, August, 96
Welby, Sir Reginald, 104, 125–6
Weldon, W. F. R., 99, 178–9
Wilks, John, 4, 11–12
Woods, Robert, 92
Working Party on Hospital Records, 213
World Health Organisation, 21, 208, 213
Wright, Maurice, 104

Yule, George Udny, 179–82, 195, 207

Zuboff, Shoshana, 177